WITHDRAWN

GLOBAL WARMING
IN THE
21ST CENTURY

GLOBAL WARMING IN THE 21ST CENTURY

VOLUME

2 Melting Ice
and Warming Seas

Bruce E. Johansen

Praeger Perspectives

Westport, Connecticut
London

Library of Congress Cataloging-in-Publication Data

Johansen, Bruce E. (Bruce Elliott), 1950–
    Global warming in the 21st century / Bruce E. Johansen.
        v.   cm.
    Includes bibliographical references and index.
    Contents: v.1. Our evolving climate crisis—v.2. Melting ice and warming
seas—v.3. Plants and animals in peril
    ISBN 0-275-98585-7 (set : alk. paper)—ISBN 0-275-98586-5
(v. 1 : alk. paper)—ISBN 0-275-98587-3 (v. 2 : alk. paper)—
ISBN 0-275-99093-1 (v. 3 : alk. paper)   1. Global warming.   2. Global
warming—Environmental aspects.   3. Renewable energy resources.
4. Global warming—Political aspects.   I. Title: Global warming in the
twenty-first century.   II. Title.
    QC981.8.G56J643   2006
    363.738'74—dc22   2006006633

British Library Cataloguing in Publication Data is available.

Library of Congress Catalog Card Number: 2006006633
ISBN: 0–275–98585–7 (set)
        0–275–98586–5 (vol.1)
        0–275–98587–3 (vol.2)
        0–275–99093–1 (vol.3)

First published in 2006

Praeger Publishers, 88 Post Road West, Westport, CT 06881
An imprint of Greenwood Publishing Group, Inc.
www.praeger.com

Printed in the United States of America

The paper used in this book complies with the
Permanent Paper Standard issued by the National
Information Standards Organization (Z39.48–1984).

10 9 8 7 6 5 4 3 2 1

CONTENTS

Contents

Color insert in Volume 2 precedes Part IV.

PREFACE

The Next Energy Revolution

In one hundred years, students of history may remark at the nature of the fears that stalled responses to climate change early in the twenty-first century. Skeptics of global warming kept change at bay, it may be noted, by appealing to most people's fear of change that might erode their comfort and employment security, all of which were psychologically wedded to the massive burning of fossil fuels. A necessary change in our energy base may have been stalled, they might conclude, beyond the point where climate change forced attention, comprehension, and action.

Technological change always generates fear of unemployment. Paradoxically, such changes also generate economic activity. A change in our basic energy paradigm during the twenty-first century will not cause the ruination of our economic base, as some skeptics of climate change believe, any more than the coming of the railroads in the nineteenth century ruined an economy in which the horse was the major land-based vehicle of transportation. The advent of mass automobile ownership early in the twentieth century propelled economic growth, as did the transformation of information gathering and handling via computers in the recent past. The same developments also put blacksmiths, keepers of hand-drawn accounting ledgers, and anyone who repaired manual typesetters out of work.

We are overdue for an energy system paradigm shift. Limited oil supply and its location in the volatile Middle East make a case for new sources, along with the accelerating climate change from greenhouse gases accumulating in the atmosphere. According to an editorial in *Business Week*, "A national policy that cuts fossil-fuel consumption converges with a geopolitical policy of reducing energy dependence on Middle East oil. Reducing carbon dioxide emissions is no longer just a 'green' thing. It makes business and foreign policy sense, as well. . . . In the end, the only real solution may be new energy technologies. There has been little innovation in energy since the internal combustion engine was invented in the 1860s and Thomas Edison built his first commercial electric generating plant in 1882" ("How to Combat Global Warming" 2004, 108).

Even the climate skeptics don't deny that climate has warmed. Temperatures have, indeed, been rising. Nine of the ten warmest years worldwide have been recorded since 1990. General warming does not imply that all cold weather has ended—instead, extremes have generally been increasing. As global averages have increased, for example, a spell of intense cold killed more than 1,000 people in India and Bangladesh during the winter of 2002–2003, only a few months after hundreds perished of record heat in the same area.

Increasing evidence also indicates that rising temperatures are changing the hydrological cycle, helping to cause intensifying chances of precipitation extremes of drought and deluge. Western Europe has experienced flooding rains while the western interior of North America has been suffering what may be the worst drought since that of a thousand years ago, which ruined the civilization of the Anasazis. Intensity of storms often increases with warmth. In the midst of drought during the summer of 2002, for example, sections of Nebraska experienced cloudbursts that eroded soil and washed out an interstate highway. Hours later, the drought returned.

Temperature also does not fully express itself immediately, but through a feedback loop of perhaps a half century. We are, thus, now experiencing the climate related to greenhouse gas levels of about 1960. Since that time, carbon dioxide levels have risen substantially, all but guaranteeing further, substantial warming during at least the next half-century.

As temperatures rise, energy policy in the United States under the George W. Bush administration generally ignores atmospheric physics. Gas mileage for U.S. internal combustion engines has, in fact, declined during the last two decades, but gains in energy efficiency have been more than offset by increases in vehicle size, notably through sport-utility vehicles. And now comes a mass advertising campaign aimed at security-minded U.S. citizens for the biggest gas-guzzler of all, the fortress-like Humvee. Present-day automotive marketing may seem quaint in a hundred years. By the end of this century, perhaps sooner, the internal combustion engine and the oil (and natural gas) burning furnace will become museum pieces. They will be as antique as the horse and buggy is today. Such change will be beneficial and necessary.

As of 2005, the federal government of the United States (which, as a nation, produces almost one-quarter of the world's greenhouse gases) was sitting out the next worldwide energy revolution. The United States is being led (if that is the word) by a group of minds still set to the clock of the early twentieth-century fossil fuel boom. The Bush administration not only has refused to endorse the Kyoto Protocol, but also has (with a few exceptions, such as its endorsement of hydrogen and hybrid-fueled automobiles) failed to take seriously the coming revolution in the technology of energy production and use. In a century, George Bush's bust may sit in a greenhouse gas museum, not far from a model of an antique internal-combustion engine. A plaque may mention his family's intimate ties to the oil industry as a factor in his refusal to think outside that particular box.

As the White House banters about "sound science," yellow-jacket wasps were sighted on Northern Baffin Island during the summer of 2004. By the end of the twenty-first century, if "business as usual" fossil fuel consumption is not curbed substantially, the atmosphere's carbon dioxide level will reach 800 to 1,000 parts per million. The last time this level was reached 55 million years ago, during the days of the dinosaurs, the water at the North Pole, then devoid of ice, reached approximately 68 degrees F.

Before the end of this century, the urgency of global warming will become manifest to everyone. Solutions to our fossil fuel dilemma—solar, wind, hydrogen, and others—will evolve during this century. Within our century, necessity will compel invention. Other technologies may develop that have not, as yet, even broached the realm of present-day science fiction any more than digitized computers had in the days of the Wright Brothers a hundred years ago. We will take this journey because the changing climate, along with our own innate curiosity and creativity, will compel a changing energy paradigm.

Such change will not take place at once. A paradigm change in basic energy technology may require the better part of a century, or longer. Several technologies will evolve together. Oil-based fuels will continue to be used for purposes that require it. (Air transport comes to mind, although engineers already are working on ways to make jet engines more efficient.)

A wide variety of solutions are being pursued around the world, of which the following are only a few examples. Some changes involve localities. Already, several U.S. states are taking actions to limit carbon dioxide emissions despite a lack of support from the U.S. federal government. Building code changes have been enacted. Wind-power incentives have been enacted—even in Bush's home state of Texas, where some oil fields now host wind turbines.

Wind turbines and photovoltaic solar cells are becoming more efficient and competitive. Improvements in farming technology are reducing emissions. Deep-sea sequestration of CO_2 is proceeding in experimental form, but with concerns about this technology's effects on ocean biota. Tokyo, where a powerful urban heat island has intensified the effects of general warming, has proposed a gigantic ocean-water cooling grid. Britain and other countries are considering carbon taxes.

J. Craig Venter, the maverick scientist who compiled a human genetic map with private money, has decided to tap a $100 million research endowment he created from his stock holdings to scour the world's deep ocean trenches for bacteria that might be able to convert carbon dioxide to solid form using little sunlight or other energy. Failing that, Venter proposes to synthesize such organisms via genetic engineering. He would like to invent two synthetic microorganisms: one to consume carbon dioxide and turn it into raw materials comprising the kinds of organic chemicals that are now made from oil and natural gas, the other to generate hydrogen fuel from water and sunshine.

The coming energy revolution will engender economic growth and become an engine of wealth creation for those who realize the opportunities that it offers. Denmark, for example, is making every family a share owner in a burgeoning wind-power industry. The United Kingdom is making plans to reduce its greenhouse gas emissions 50 percent in 50 years. The British program begins to address the position of the Intergovernmental Panel on Climate Change (IPCC) that emissions will have to fall 60 to 70 percent by century's end to avoid significant warming of the lower atmosphere due to human activities. The Kyoto Protocol, with its reductions of 5 to 15 percent (depending on the country) is barely earnest money compared to the required paradigm change, which will reconstruct the system and the way most of the world's people obtain and use energy.

Solutions will combine scientific achievement and political change. We will end this century with a new energy system, one that acknowledges nature and works with its needs and cycles. Economic development will become congruent with the requirements of sustaining nature. Coming generations will be able to mitigate the effects of greenhouse gases without the increase in poverty so feared by skeptics. Within decades, a new energy paradigm will be enriching us and securing a future that works with the requirements of nature, not against it.

So, how much "wiggle room" does the Earth and its inhabitants have before global warming becomes a truly world-girdling disaster, rather than what many people take to be a set of political, economic, and scientific debating points? Perhaps the best synopsis was provided by James Hansen during a presentation at the American Geophysical Union annual meeting in San Francisco, December 6, 2005. The Earth's temperature, with rapid global warming over the past 30 years, said Hansen, is now passing through the peak level of the Holocene, a period of relatively stable climate that has existed for more than 10,000 years. Further warming of more than 1 degree C "will make the Earth warmer than it has been in a million years. 'Business-as-usual' scenarios, with fossil fuel CO_2 emissions continuing to increase at about 2 percent a year as in the past decade, yield additional warming of 2 or 3 degrees C this century and imply changes that constitute practically a different planet" (Hansen 2005).

Stop for a moment, and ponder the words, delivered in the measured tones of a veteran scientist: "Practically a different planet," a very real probability by the end of the twenty-first century. Hansen is not joking. He continued: "I present multiple lines of evidence indicating that the Earth's climate is nearing, but has not passed, a tipping point, beyond which it will be impossible to avoid climate change with far-ranging undesirable consequences" (Hansen 2005).

Coming to cases, Hansen described changes that will include:

not only loss of the Arctic as we know it, with all that implies for wildlife and indigenous peoples, but losses on a much vaster scale due to worldwide rising seas. Sea level will increase slowly at first, as losses at the fringes of Greenland and Antarctica due to accel-erating ice streams are nearly balanced by increased snowfall and ice sheet thickening in the ice sheet interiors. But as Greenland and West Antarctic ice is softened and lubricated by melt-water and as buttressing ice shelves disappear due to a warming ocean, the balance will tip toward ice loss, thus bringing multiple positive feedbacks into play and causing rapid ice sheet disintegration. The Earth's history suggests that with warming of 2 to 3 degrees C the new equilibrium sea level will include not only most of the ice from Greenland and West Antarctica, but a portion of East Ant-arctica, raising sea level of the order of 25 meters (80 feet).

To be judicious—we don't want to ruin our case with over-statement—one might allow perhaps two or three centuries for a tem-perature rise in the atmosphere to express itself as sea-level rise from melting ice. Contrary to lethargic ice sheet models, Hansen suggests, real-world data suggest substantial ice sheet and sea-level change in centuries, not millennia. Now take a look at a map of the world and pay attention to the coastal urban areas. Is anyone worried yet?

Hansen hopes that "the grim 'business-as-usual' climate change" may be avoided by slowing the growth of greenhouse gas emissions during the first quarter of the present century, requiring "strong policy leadership and international cooperation" (Hansen 2005). However, he noted (venturing into the realm of politics) that "special interests have been a roadblock wielding undue influence over policymakers. The special interests seek to maintain short-term profits with little regard to either the long-term impact on the planet that will be inherited by our children and grandchildren or the long-term economic well-being of our country" (Hansen 2005). Hansen leaves to the audience the task of putting names and faces to the special interests who, along with the rest of us, are attending this crucial juncture in the history of the planet and its inhabitants.

FURTHER READING

Hansen, James E. Is There Still Time to Avoid "Dangerous Anthropogenic Interference" with Global Climate? A Tribute to Charles David Keeling. Paper delivered to the American Geophysical Union, San Francisco, December 6, 2005. www.columbia.edu/~jeh1/keeling_talk_and_slides.pdf.

"How to Combat Global Warming: In the End, the Only Real Solution May Be New Energy Technologies," *Business Week*, August 16, 2004, 108.

ACKNOWLEDGMENTS

Anyone who has written and published a book knows well that it is hardly a solitary journey, even after many hundreds of hours alone at the keyboard. Along the way, many thanks are due, in my case to my wife Pat Keiffer and my family (Shannon, Samantha, Madison), who kept me clothed and fed while enduring numerous bulletins from global warming's many scientific and political fronts. Gratitude also is due to the people of University of Nebraska at Omaha Interlibrary Loan, who can get just about anything that's been published anywhere; to my editors Heather Staines and Lisa Pierce; to the production crew; to the University of Nebraska at Omaha School of Communication Director Jeremy Lipschultz (himself an accomplished author who knows my working habits); and Deans Robert Welk and Shelton Hendricks (for partial relief from teaching duties). Further debts are owed to manuscript reviewers Andrew Lacis of NASA's Goddard Institute for Space Studies in New York City; Gian-Reto Walther of the Institute of Geobotany, University of Hannover, Germany; and Julienne Stroeve, with the Cooperative Institute for Research in Environmental Studies, University of Colorado, Boulder.

INTRODUCTION: VOLUME 2

Approaching the Point of Irreversible Meltdown

Among scientists who keep tabs on the pace of global warming, anxiety has been rising that the Earth is reaching an ominous threshold, a point of no return ("tipping point" in some of the scientific literature). Within a decade or two, various feedback mechanisms will accelerate the pace of greenhouse warming past any human ability to contain or reverse it. Carbon dioxide levels in the atmosphere are rising rapidly, fed, among other provocations, by increasing fossil fuel use in the United States, melting permafrost, slash-and-burn agriculture in Indonesia and Brazil, increasing wildfires, as well as rapid industrialization using dirty coal in China and India.

All of this is taking place amid an air of fossil-fueled complacency in U.S. halls of power, where global warming has been ignored. Sir John Houghton, one of the world's leading experts on global warming, told the *London Independent*, "We are getting almost to the point of irreversible meltdown, and will pass it soon if we are not careful" (Lean 2004, 8). Volume 2 of this set considers the evidentiary trail written in melting ice (Arctic, Antarctic, and mountain glaciers) and in warming seas and discusses their effects on plant and animal life in the oceans that cover two-thirds of the Earth's surface.

The evidence of cascading climate change is most dramatic in the Arctic, described in Chapter 10. Inuit were surprised during the summer of 2004 by the arrival of several *Vespula intermedia* (yellow-jacket wasps) in Arctic Bay, a community of 700 people on the northern tip of

Baffin Island at more than 73 degrees north latitude. Noire Ikalukjuaq, the mayor of Arctic Bay, photographed one of the wasps at the end of August. Ikalukjuaq said he knows no word in Inuktitut, the Inuit language, for the wasps, indicating that these people have never seen this species before. Other people in the same community told him they had seen wasps at about the same time.

During the summer of 2004, enough additional Arctic ice to blanket Texas twice over melted as compared to the previous year. In the past, weak ice years often were followed by heavy ice years, when cold winters or cool summers maintained or extended the icepack. This balance, however, has not been occurring recently. "If you look at these last few years, the loss of ice we've seen . . . is rather remarkable," Mark Serreze of the National Snow and Ice Data Center at the University of Colorado told Katy Human of the *Denver Post* (Human 2004, B-2). The year 2004 was the third year in a row with extreme ice losses, pointing to an acceleration of the downward trend, he said.

Addressing a Senate Commerce Committee hearing on global warming August 15, 2004, Sheila Watt-Cloutier, president of the Inuit Circumpolar Conference, said: "The Earth is literally melting. If we can reverse the emissions of greenhouse gases in time to save the Arctic, then we can spare untold suffering." She continued, "Protect the Arctic and you will save the planet. Use us as your early-warning system. Use the Inuit story as a vehicle to reconnect us all so that we can understand the people and the planet are one." The Inuits' ancient connection to their hunting culture may disappear within her grandson's lifetime, Watt-Cloutier said. "My Arctic homeland is now the health barometer for the planet" (Pegg 2004).

The changing nature of weather in the far northern latitudes is evident across the region: Spruce-bark beetles, their reproductive cycles accelerating as the weather warms, have destroyed forests across large regions of Alaska's Kenai Peninsula, as the Alaskan village of Shishmaref erodes into a winter sea no longer contained by ice. Ice at the edges of Greenland slides into the sea with unaccustomed speed as Russian peat bogs smolder. Some hockey teams in northern Canada have imported equipment to freeze their rinks, which now melt before their playing seasons end.

The pace at which ice melts has become the pulse of a warming planet. On the Antarctic Peninsula (Chapter 11), scientists study ways in which the collapse of coastal ice may accelerate the movement of interior glaciers toward the seas. Grass sprouts in areas of the Antarctic Peninsula that heretofore had been covered by ice most of the year. Meanwhile, as described in Chapter 12, glaciers retreat in mountains worldwide, to the point where Glacier National Park has lost most of its namesake ice and Hemingway's snows of Kilimanjaro may be gone within twenty years. Climbers are plucked from the Alps as parts of melting mountains collapse under their feet.

As ice melts, seas rise from the infusion of fresh water as well as thermal expansion, the subjects of Chapters 13–15. In some areas, such as the U.S. Atlantic and Gulf of Mexico coasts, the net rise in the seas is being accelerated by natural subsidence. Natural subsidence itself is being sped in the same places by the withdrawal of underground water (and, in some cases, oil) for human uses. The rising sea levels have threatened many marshlands and their associated webs of life.

In the United Kingdom, "managed realignment" has begun along some coastal areas as seawater is being allowed to reclaim farmland and marshes. Some of the links at the St. Andrews golf course, where golf was born, have eroded into the ocean. In London, government bodies have been commissioned to study the possibility that the United Kingdom's seat of government may be relocated as rising seas swamp the Thames River. The residents of Venice, Italy, debate whether to surrender to rising waters or build an expensive gating system that might hold the sea for a few more decades. Low-lying areas around the world, from Bangladesh to small island nations, face the day when the sea may inundate their homelands.

As seas rise, scientists have realized that warming's effects on the oceans involve much more than the swamping of coastlines, as important as these effects may be. The oceans' grand conveyor belt, the thermohaline circulation, may be imperiled as cold, fresh water from melting ice invades the oceans. In the meantime, rising carbon dioxide levels in the oceans are affecting marine life.

As the twenty-first century dawned, carbon dioxide levels were rising in the oceans more rapidly than at any time since the age of the dinosaurs, according to a report published September 25, 2003, in *Nature*. Oceanographers Ken Caldeira and Michael E. Wickett wrote: "We find that oceanic absorption of CO_2 from fossil fuels may result in larger pH changes over the next several centuries than any inferred in the geological record of the last 300 million years, with the possible exception of those resulting from rare, extreme events such as bolide impacts or catastrophic methane hydrate degassing" (Caldeira and Wickett 2003, 365). A "bolide" is a large extraterrestrial body (usually at least a half mile in diameter, sometimes much larger) that impacts the Earth at a speed roughly equal to that of a bullet in flight.

Until now, some climate experts have asserted that the oceans would help control the rise in atmospheric carbon dioxide by acting as a filter. Various commercial interests have proposed impregnating the ocean depths with our unwanted surplus carbon. Caldeira and Michael Wickett assert, however, that carbon dioxide that is removed from the atmosphere enters the oceans as carbonic acid, gradually altering the acidity of ocean water. According to their studies, the change over the last century already matches the magnitude of the change that occurred in the entire 10,000 years preceding the industrial age. Caldeira pointed to acid rain from industrial emissions as a possible precursor of changes in the oceans. "Most ocean life resides near the surface, where the greatest change would be expected to come, but deep ocean life may prove to be even more sensitive to changes," Caldeira said (Toner 2003).

Marine plankton and other organisms whose skeletons or shells contain calcium carbonate, which dissolves with rising acidity, may be particularly vulnerable. Coral reefs—already suffering from pollution, rising ocean temperatures, and other stresses—are comprised almost entirely of calcium carbonate. "It's difficult to predict what will happen because we haven't really studied the range of impacts," Caldeira said. "But we can say that if we continue business as usual, we are going to see some significant changes in the acidity of the world's oceans" (Toner 2003).

Global warming is contributing to changes that some scientists call an "ecological meltdown," with devastating implications for fisheries and wildlife. The meltdown begins at the base of the food chain as rising ocean temperatures kill plankton or force them toward the poles, affecting fish stocks and sea-bird populations.

Scientists at the Sir Alistair Hardy Foundation for Ocean Science in Plymouth, England, which has been monitoring plankton growth in the North Sea for more than seventy years, have stated that an unprecedented warming of the North Sea has driven plankton hundreds of miles to the north. The plankton have been replaced by smaller, warm-water species that are less nutritious (Sadler and Lean 2003, 12). Although overfishing of cod and other species has played a role, fish stocks have not recovered after cuts in fishing quotas. The number of salmon returning to British waters is now half what it was twenty years ago, and a decline in plankton is a major factor.

Research by the British Royal Society for the Protection of Birds (RSPB) has established that sea-bird colonies off the Yorkshire coast and the Shetlands during 2003 "suffered their worst breeding season since records began, with many simply abandoning nesting sites" (Sadler and Lean 2003, 12). These sea birds are declining in numbers because the sand eels they eat are dying. The sand eels feed on plankton. This survey concentrated on kittiwakes, one breed of sea birds, but other species that feed on the eels, including puffins and razorbills, have also been seriously affected. Euan Dunn of the RSPB commented, "We know that sand eel populations fluctuate and you do get bad years. But there is a suggestion that we are getting a series of bad years, and that suggests something more sinister is happening" (p. 12).

FURTHER READING

Caldeira, Ken, and Michael E. Wickett. "Oceanography: Anthropogenic Carbon and Ocean pH." *Nature* 425 (September 25, 2003): 365.

Human, Katy. "Disappearing Arctic Ice Chills Scientists: A University of Colorado Expert on Ice Worries That the Massive Melting Will Trigger

Dramatic Changes in the World's Weather." *Denver Post*, October 5, 2004, B-2.

Lean, Geoffrey. "Global Warming Will Redraw Map of the World." *London Independent*, November 7, 2004, 8.

Pegg, J. R. "The Earth Is Melting, Arctic Native Leader Warns." Environment News Service, September 16, 2004.

Sadler, Richard, and Geoffrey Lean. "North Sea Faces Collapse of Its Ecosystem." *London Independent*, October 19, 2003, 12.

Toner, Mike. "Oceans' Acidity Worries Experts. Report: Carbon Dioxide on Rise, Marine Life at Risk." *Atlanta Journal and Constitution*, September 25, 2003. (Lexis).

III ICEMELT AROUND THE WORLD

INTRODUCTION

The most widespread indication that the Earth is steadily warming has been the steady erosion of ice in the Arctic, Antarctic, and on mountain glaciers. Although a few exceptions do exist, the worldwide erosion of ice leaves little doubt that the Earth has experienced steady warming for at least a century. The causes of this warming are still open to debate: is it caused by human activity, a product of changes in atmospheric circulation, or both? The following global survey indicates that icemelt is accelerating. The melting of ice has profound implications not only for Arctic, Antarctic, and mountain ecosystems but also for hundreds of millions of people living at lower elevations who depend upon glacier melt for water and electricity generation. Billions more around the Earth who live on or near continental coasts and islands have (and will) feel the effects of global icemelt through gradually rising sea levels.

10 CLIMATE CHANGE IN THE ARCTIC

Warming is being felt most intensely in the Arctic, where a world based on ice and snow has been melting away. Arctic sea-ice cover shrank more dramatically between 2002 and 2004 than at any time since detailed records have been kept. A report produced by 250 scientists under the auspices of the Arctic Council found that Arctic sea ice was half as thick in 2003 as it was thirty years earlier. If present rates of melting continue, there may be no summer ice in the Arctic by 2070, according to the study. Pal Prestrud, vice chairman of the steering committee for the report, said, "Climate change is not just about the future; it is happening now. The Arctic is warming at twice the global rate" (Harvey 2004, 1).

During the summer of 2004, enough Arctic ice to blanket Texas twice over advected or compacted. In the past, years with a relatively large amount of ice melt often were followed by recovery the next year, when cold winters or cool summers kept ice from melting. That kind of balancing cycle stopped after 2002. "If you look at these last few years, the loss of ice we've seen, well, the decline is rather remarkable," said Mark Serreze of the National Snow and Ice Data Center at the University of Colorado (Boyd 2002, A-1; Human 2004, B-2). The year 2004 was the third year in a row with extreme ice losses, indicating an acceleration of the melting trend. Arctic ice has been declining about 8 percent per decade, and the trend is accelerating.

At the same time, Alaska's boreal forests are expanding northward at a rate of about 100 kilometers per 1 degree C rise in temperatures. Ice cover on lakes and rivers in the mid to high northern latitudes now lasts for about two weeks less than it did 150 years ago. During late June 2004, as temperatures in Fairbanks, Alaska, peaked in the upper eighties, a flood warning was issued near Juneau, Alaska—not due to rainfall, but because of glacial snowmelt.

German scientists said late in 2004 that they had detected a major temperature rise in the Arctic Ocean that they associated with pro-gressive shrinking of the region's sea ice. Temperatures recorded in the upper 500 meters (1,625 feet) of sea in the Fram Strait—the gap between Greenland and the Norwegian island of Spitsbergen—were 0.6 degrees C (1.08 degrees F) higher than in 2003. The temperature rise was de-tectable to a water depth of 2,000 meters (6,500 feet), "representing an exceptionally strong signal by ocean standards." Scientists from the Alfred Wegener Institute for Polar and Marine Research in Bre-merhaven stated that water in the Fram Strait has been warming steadily since 1990; during the past three years, satellite images have documented "a clear recession" of sea-ice edges, both in the strait and in the Barents Sea ("Major Temperature Rise" 2004).

PERSONAL STORIES OF CLIMATE CHANGE

Rapid climate change in the Arctic has long since passed the point of speculation. The destruction of an ecosystem heretofore based on ice and snow is now a day-to-day reality in the lives of people who live near or above the Arctic Circle. Their personal stories indicate that the atmosphere is warming more rapidly in parts of the Arctic than any-where else on Earth.

Around the Arctic, in Inuit villages now connected by e-mail as well as the oral history of traveling hunters, weather watchers are reporting striking evidence that global warming is an unmistakable reality. Weather reports from the Arctic sometimes read like a tape of the Intergovern-mental Panel on Climate Change (IPCC) on fast forward. These personal stories support IPCC expectations that climate change will be felt most dramatically in the Arctic.

Addressing a Senate Commerce Committee hearing on global warming on August 15, 2004, Sheila Watt-Cloutier, president of the Inuit Circumpolar Conference, said: "The Earth is literally melting. If we can reverse the emissions of greenhouse gases in time to save the Arctic, then we can spare untold suffering." She continued, "Protect the Arctic and you will save the planet. Use us as your early-warning system. Use the Inuit story as a vehicle to reconnect us all so that we can understand the people and the planet are one" (Pegg 2004). The Inuits' ancient connection to their hunting culture may disappear within her grandson's lifetime, Watt-Cloutier said. "My Arctic

Sheila Watt-Cloutier. Courtesy of Stephen Hendrie/Inuit Circumpolar Conference.

homeland is now the health barometer for the planet" (2004). Committee chair John McCain, an Arizona Republican, said a recent trip to the Arctic showed him that "these impacts are real and consistent with earlier scientific projects that the Arctic region would experience the impacts of climate change at a faster rate than the rest of the world. We are the first generation to influence the climate and the last generation to escape the consequences," McCain said (2004).

Alaskan natives have established a Web site (nativeknowledge.org) for sharing their experiences with climate change: "Turtles appearing for the first time on Kodiak Island, birds starving on St. Lawrence Island, thunder first heard on Little Diomede Island, ... snowmobiles falling through the ice in Nenana.... Already the central Arctic is warming 10 times as fast as the rest of the planet, outpacing even our attempts to describe it" (Frey 2002, 26).

Sachs Harbour, a village on Banks Island above the Arctic Circle, is sinking into the permafrost as its 130 residents swat mosquitoes. Summer downpours of rain with thunder, hail, and lightning have swept over Arctic islands for the first time in anyone's memory. Swallows, sand flies, robins, and pine pollen are being seen and experienced by people who have never known them. Shishmaref, an Inuit village on the far western lip of Alaska sixty miles north of Nome, is being washed into the newly liquid (and often stormy) Arctic Ocean as its permafrost base dissolves.

During the summer of 2004, several *Vespula intermedia* (yellow-jacket wasps) were sighted in Arctic Bay, a community of 700 people on the northern tip of Baffin Island at more than 73 degrees north latitude. Noire Ikalukjuaq, the mayor of Arctic Bay, photographed one of the wasps at the end of August. Ikalukjuaq, who said he knew no word in Inuktitut (the Inuits' language) for the insect, reported that other people in the community had also seen wasps at about the same time ("Rare Sighting" 2004).

While wasps were being observed on northern Baffin Island, blue mussels were found growing on seabeds about 1,300 kilometers from the North Pole. The blue mussels, which usually favor warmer waters such as those off France or the eastern United States, were discovered during August 2004 off Norway's Svalbard Archipelago in waters covered with ice most of the year. "The climate is changing fast," said Geir Johnsen, a professor at the Norwegian University for Science and Technology who was among the experts who found the bivalves. Mollusks are a "very good indicator that the climate is warming," he added. "It seems like the mussels we found are two to three years old" ("North Pole Mussels" 2004, A-16).

In the Eskimo village of Kaktovik, Alaska, on the Arctic Ocean roughly 250 miles north of the Arctic Circle, a robin built a nest during 2003—not an unusual event in more temperate latitudes but quite a departure from the usual in a place where, in the Inupiat Eskimo language, no name exists for robins. In the Okpilak River valley, which has heretofore been too cold and dry for willows, they are sprouting profusely; never mind that, in the Inupiat language, "Okpilak" means "river with no willows" (Kristof 2003). Three kinds of salmon have been caught in nearby waters in places where they were once unknown.

Correspondent Jerry Bowen, appearing on the CBS Morning News August 29, 2002, from Barrow, Alaska, quoted Simeon Patkotak, a native elder, as saying that residents there had just witnessed their first mosquitoes. Ice cellars carved out of permafrost were melting as well, forcing local native people to borrow space in electric freezers for the first time to store whale meat. Average temperatures in Barrow have risen 4 degrees F during the past thirty years (Bowen 2002). The average date at which the last snow melts at Barrow in the spring or summer has receded about forty days between 1940 and the years after 2000, from early July to (some years) as early as mid- or late May (Wohlforth 2004, 27).

Temperatures in the Arctic rose so rapidly during the spring of 2004 that Ben Saunders, aged twenty-six, was forced to abandon his attempt to become the first person to trek solo across the North Pole from Russia to Canada. Saunders reached the North Pole May 11 but had to be rescued three days later, about fifty kilometers on the Canadian side, when he hit open water. He was airlifted to Ottawa. Mr. Saunders, from London, began his journey at Cape Artichevsky in northern Siberia on March 5. His journey ended seventy-two days and 965.22 kilometers later. "The weather this year was the warmest since they began keeping records," Saunders said. "And the sea ice coverage in the Arctic last year was the least ever, according to NASA. They haven't calculated it yet for this year, but we expect the sea ice coverage to be even lower" (Kristal-Schroder 2004, A-1).

Saunders said that during some days of his trek the temperature rose to 15 degrees C, as he skied without a hat or mittens. Constant thawing and freezing cycles made the ice especially hazardous. "I've never seen myself as an eco-warrior," Saunders wrote on his Web site a few days before reaching the pole. "I've been wary of taking a stance on climate change, as I don't believe we know enough about what's going on, but it's obvious that things are changing fast. It's an issue I'll certainly be taking far more interest in" (Kristal-Schroder 2004, A-1).

TEMPERATURES RISE IN THE ARCTIC

Several data sources, including those from Russian North Pole measurements (1950–1991), manned drifting camps, arrays of buoys, and satellite retrievals, indicate an increase in Arctic Ocean surface air temperature (SAT). Wintertime SAT anomalies from the average over the Arctic and the mid-latitudes indicate two characteristic warming events: the first during the mid-1920s to 1940, and the second after 1980 (Semenov and Bengtsson 2003). The warming during the first part of the twentieth century had its largest amplitude in the Arctic above 60 degrees north latitude. The warming today is more widespread than that of the 1920s and 1930s.

Josefino Comiso of NASA's Goddard Space Flight Center in Greenbelt, Maryland, compiled surface temperature measurements from several points in the Arctic between 1981 and 2001 using satellite thermal infrared data. Large warming anomalies were found over sea ice, Eurasia, and North America, but temperatures in Greenland varied from no change to a slight cooling in some areas. Temperature increases were more rapid during the 1990s than in the 1980s. Warming lengthened the melt season (when temperatures are above the freezing point) by ten to seventeen days, and the rate of warming between 1981 and 2001 was eight times as rapid as during the previous century (Comiso 2002, 1956, and 2003, 3498).

"There may be a natural part of it, but there's something else being superimposed on top of it," said glaciologist Lonnie Thompson about the accelerating Arctic icemelt indicated by Comiso's data. "And it matches so many other lines of evidence of warming. Whether you're talking about bore-hole temperatures, shrinking Arctic sea ice, or glaciers, they're telling the same story" (Revkin 2001, A-1).

According to NASA satellite surveys, perennial (year-round) sea ice in the Arctic has been declining at a rate of 9 percent per decade (Stroeve et al. 2005). During 2002, summer sea ice was at record low levels, a trend that persisted in 2003 and 2004 ("Recent Warming" 2003; Stroeve et al. 2005). During the past thirty-five years, Arctic sea ice also thinned by more than 40 percent—from an average of nine feet to about five feet thick (Toner 2003, 1-A). Research by scientists at University College London and the British Met Office's Hadley Centre for Climate Prediction and Research indicates that Arctic ice thinned from 3.5 meters (11.5 feet) thirty years ago to less than 2 meters in 2003. By the middle of the twenty-first century, according to NASA projections, the Arctic could be ice-free during the summer months (Comiso 2003, 3498).

SURFACE ALBEDO AND OTHER NATURAL ACCELERATORS OF WARMING

The melting of ocean-borne ice in polar regions can accelerate overall warming as it changes surface albedo, or reflectivity. The darker a surface, the more solar energy it absorbs. Sea water absorbs 90–95 percent of incoming solar radiation, whereas snow-free sea ice absorbs only 60–70 percent of solar energy. If the sea ice is covered with snow, the amount of absorbed solar energy decreases substantially, to only 10–20 percent. As the oceans warm and the snow and ice melt, therefore, more solar energy is absorbed, leading to even more melting. "It is feeding on itself now, and this feedback mechanism is actually accelerating the decrease in sea ice," said Mark Serreze of the University of Colorado (Toner 2003, 1-A).

Albedo seems a simple process, but the gradual darkening of Arctic surfaces can produce significant changes in the amount of solar energy the area absorbs. Scientists have been trying to model how large this effect may be, and by 2005, their estimates were quite substantial: three watts per square meter *per decade*, or roughly equal to a doubling of atmospheric carbon dioxide in areas where albedo has changed to a marked degree. "The continuation of current trends in shrub and tree expansion could further amplify this atmospheric heating by two to seven times," the same study anticipated (Chapin et al. 2005, 657). Forest is generally darker in color than the tundra that it replaces.

Warming recently has been shortening the snow-covered season by roughly 2.5 days per decade in the Arctic, accelerating the change in albedo (Foley 2005, 627).

Changes in reflectivity or albedo (Latin for "whiteness") are among the factors contributing to a rate of warming in the Arctic during the last twenty years that has been eight times the rate of warming during the previous 100 years ("Recent Warming" 2003). Recent increases in the number and extent of boreal forest fires have been adding to the amount of soot in the atmosphere, which changes albedo as well.

As high latitudes warm and the coverage of sea ice declines, thawing Arctic soils may release significant amounts of carbon dioxide and methane now trapped in permafrost. Warmer ocean waters also could release formerly solid methane and carbon dioxide from the sea floor. According to David Rind of NASA's Goddard Institute for Space Studies in New York, "these feedbacks are complex and we are working to understand them. Global warming is usually viewed as something that's 50 or 100 years in the future, but we have evidence that the climate of the Arctic is changing right now, and changing rapidly. Whatever is causing it, we are going to have to start adapting to it" (Toner 2003, 1-A).

The warming of the Arctic "will definitely impact our weather in the United States," added Rind. "Those outbreaks of Arctic cold we get each winter might seem like something we could do without, but if we don't have them, we're going to receive a lot less winter precipitation," he said. "Computer models show that Kansas would be 10 degrees warmer during the winter—and get 40 per cent less snow, which could make it very difficult to grow winter wheat there" (Toner 2003, 1-A).

DRAMATIC WARMING IN THE CANADIAN ARCTIC: A SLICE OF DAILY LIFE

Global warming has become a fact of daily life in the Canadian Arctic. July 29, 2001, for example, was a very warm day in Iqaluit, capital of the new semi-sovereign Inuit nation of Nunavut in the Canadian Arctic. The bizarre weather is the talk of the town. The urgency of global warming is on everyone's lips. As U.S. President George W. Bush fretted about "sound science," the temperature hit 25 degrees C on July 28 in this Baffin Island community that nudges the Arctic Circle, following a string of days that were nearly as warm. It was the warmest summer anyone in the area could remember. Travelers joked about forgetting their shorts, sunscreen, and mosquito repellant— all now necessary equipment for a globally warmed Arctic summer.

In Iqaluit, a warm, desiccating northwesterly wind raised whitecaps on nearby Frobisher Bay and rustled carpets of purple saxifrage flowers as people emerged from their overheated houses (which have been built to absorb every scrap of passive solar energy), swabbing ice cubes wrapped in hand towels across their foreheads. The high temperature, at 78 degrees F, was 30 degrees above the July average of 48, comparable to a 110- to 115-degree day in New York City or Chicago. The wind raised eddies of dust on Iqaluit's gravel roads as residents swatted slow, corpulent mosquitoes.

In Iqaluit (pronounced "Eehalooeet"), people speak of a natural world that is being turned upside down in ways that might startle even George W. Bush. "We have never seen anything like this. It's scary, *very* scary," said Ben Kovic, Nunavut's chief wildlife manager. "It's not every summer that we run around in our T-shirts for weeks at a time" (Johansen 2001, 18). At 11:30 a.m. on a Saturday in late July, with many hours of Baffin Island's eighteen-hour July daylight remaining, Kovic was sitting in his backyard, repairing his fishing boat and wearing a T-shirt and blue jeans in the warm wind. On a nearby beach, Inuit children were building sandcastles with plastic shovels and buckets, occasionally dipping their toes in the still-frigid seawater.

Warming has extended to all seasons. In Iqaluit, for example, thunder and lightning used to be an extreme rarity. Thunderstorms are now much more common across the Arctic; the day the high temperature hit 78 degrees F in Iqaluit, the forecast for Yellowknife, in the Northwest Territories, called for a high of 85 degrees F with scattered thunderstorms. During the summer of 2004, highs in the 80s F became routine in Fairbanks, Alaska. The winters of 2000–2001 and 2001–2002 in Iqaluit were notable for liquid precipitation (freezing rain) in December. Tromso, in far northern Norway, had very little snow during the winter of 2001–2002, along with very mild weather—as warm as 15 degrees C at one point shortly after Christmas. There was almost no snow at the time. Around the Tromso airport, which is near the ocean, residents could mow their lawns (personal communication from Floyd Rudmin, Tromso, Norway, January 17, 2002).

Early in January 2004, Sheila Watt-Cloutier wrote from Iqaluit, on Baffin Island, that Frobisher Bay had just frozen over for the season at a record late date:

> We are finally into very "brrrrrr" seasonal weather and the Bay is finally freezing straight across. At Christmas time the Bay was still open and as a result of the floe edge being so close we had a family of Polar bears come to visit the town a couple of times.
>
> Also in Pangnirtung [north of Iqaluit] families from the outpost camps came into town for Christmas by boat! Imagine that the ice was not frozen at all there by Christmas. But this week our temperatures with the wind-chill reached minus 52 so we are pleased. (personal communication from Sheila Watt-Cloutier, Iqaluit, Nunavut, January 4, 2004)

Watt-Cloutier was looking out at the waters of Frobisher Bay two weeks after she had represented the Inuit at a Conference of Parties to the U.N. Framework Convention on Climate Change in Milan, Italy, where she said, in part:

> Talk to hunters across the North and they will tell you the same story—the weather is increasingly unpredictable. The look and feel of the land is different. The sea-ice is changing. Hunters are having difficulty navigating and traveling safely. We have even lost experienced hunters through the ice in areas that, traditionally, were safe! Our Premier, Paul Okalik, lost his nephew when he was swept away by a torrent that used to be a small stream. The melting of our glaciers in summer is now such that it is dangerous for us to get to many of our traditional hunting and harvesting

places. . . . Inuit hunters and elders have for years reported changes to the environment that are now supported by American, British and European computermodels that conclude climate change is amplified in high latitudes. (Watt-Cloutier 2003)

Snows in Iqaluit also now tend to be heavier and wetter than previously. Winter cold spells, which still occur, have generally become shorter, according to longtime residents of the area.

As a wildlife officer, Kovic sees changes that alarm him. Polar bears, for example, are often becoming shore dwellers rather than ice dwellers, sometimes with dire consequences for unwary tourists. The harbor ice at Iqaluit did not form in 2000 and 2001 until late November or December, five or six weeks later than usual. The ice also breaks up earlier in the spring—sometimes in May in places that once were icebound into early July. In Resolute Bay, bears attracted by the smell of seal meat have been known to chase children on their way to school. Some local Inuit hunters have turned profits in the thousands of dollars by leading tourists on hunts for landbound, hungry bears that local people have come to regard as a dangerous nuisance.

The Arctic's rapid thaw has made hunting, never a safe or easy way of life, even more difficult and dangerous. Hunters in and around Iqaluit said that the weather has been seriously out of whack since roughly the middle 1990s. Simon Nattaq, an Inuit hunter, fell through unusually thin ice and became mired in icy water long enough to lose both his legs to hypothermia, one of several injuries and deaths reported around the Arctic recently due to thinning ice.

Pitseolak Alainga, another Iqaluit-based hunter, mentioned that climate change compels caution. One must never hunt alone, he said (Nattaq had been hunting by himself). Before venturing onto ice in fall or spring, hunters should test its stability with a harpoon, he said. Alainga knows the value of safety on the water. His father and five other men died during late October 1994 after an unexpected ice storm swamped their hunting boat. The younger Alainga and one other companion barely escaped death in the same storm. He believes that more hunters are suffering injuries not only because of climate change but also because basic survival skills are not being passed from generation to generation as in years past, when most Inuit lived off the land.

Urbanization, in fact, has caused many Inuit to lose their cultural bearings as well as their hunting skills. Less than fifty years ago, suicide was virtually unknown among the Inuit of Canada's Arctic. Now they are killing themselves at a rate six times the Canadian national average. Within two or three generations, many Inuit have become urbanized as the tendrils of industrial life infiltrate the Arctic. Where visitors once arrived by dog sled or sailing ship, they now stream into Iqaluit's small but very busy airport on Boeing 727s, in which half the passenger cabin has been sequestered for freight. With no land-surface connections to the outside, freight as large as automobiles is sometimes shipped to Iqaluit by air.

The population of Iqaluit jumped from about 3,500 to 6,500 in less than three years between 1998 and 2001. Substantial suburban-style houses with mortgages of hundreds of thousands of dollars sprang up around town, rising on stakes sunk into the permafrost and granite hillsides. In other areas, ranks of walk-up apartments marched along the high ridges above Frobisher Bay. Every ounce of building material had been imported from thousands of miles away. Inuit have been moving into the town from the backcountry. The town is awash in a sea of children.

People in Iqaluit subscribe to the same television services (as well as Internet) available in "the South." Bart Simpson and Oprah Winfrey are well-known personages in Iqaluit, where some homes have sprouted satellite dishes. Iqaluit also now hosts a large supermarket of a size that matches stores in larger urban areas, except that the prices are three to four times higher than in Ottawa or Omaha. If one can afford the bill, however, orange-mango-grapefruit juice and ready-cooked BBQ buffalo wings as well as many other items of standard "southern" fare are readily available.

Climate change has been rapid and easily detectable within a single human lifetime. "When I was a child," said Watt-Cloutier, "we never swam in the river [Kuujjuaq] where I was born [Nunavik, northern Quebec]. Now kids swim there all the time in the summer." Cloutier does not remember wearing short pants as a child (Johansen 2001, 19).

Some of the rivers to which Arctic char return for spawning have dried up, according to Kovic; the summer of 2001 brought drought, as well as record heat, to Iqaluit and its hinterland. Flying above glaciers in the area, Kovic has noticed that their coloration was changing. "The glaciers are turning brown," he said, speculating that melting ice may be exposing debris and that air pollution from southern latitudes may be a factor. Some ringed seals with little or no hair have been caught, said Kovic. Asked why seals have been losing their hair, Kovic answered: "That is a big question that someone has to answer" (Johansen 2001, 19).

Climate change has become a lively issue in Inuit politics. At a press conference that was televised across Canada August 1, 2002, for example, Nunavut Premier Paul Okalik publically confronted Alberta Premier Ralph Klein after Klein warned that the terms of the Kyoto agreement could reduce oil-rich Alberta's equalization contributions to have-not regions of the country. "You can keep your money," Okalik told Klein, as he argued that global warming presents a direct threat to the Inuit way of life (Bell 2002). Alberta's government has complained that compliance with the protocol could throw thousands of Albertans out of work and take billions of dollars out of its province's economy. Klein has also contended that the U.S. refusal to ratify Kyoto would make it difficult for Alberta to compete with U.S. oil producers.

WARMING IN NORTHERNMOST CANADA

Along the shore of the Beaufort Sea, pounding surf has been eating away at the melting shoreline of Tuktoyaktuk, an Inuit town of 1,000 people at the mouth of the Mackenzie delta. In a town where explorers in 1911 found temperatures of minus 60 degrees F (with wind chill, minus 110 degrees F), residents now worry about warming that may cause the permafrost base to dissolve "like salt in the sea" (D. L. Brown 2003, A-14). Ken Johnson, a planner for an engineering company that has studied Tuktoyaktuk, said, "There is some speculation the storms off the Beaufort Sea are created by the decreasing ice pack in the Canadian north. It is creating a larger fetch, or distance of open water, that allows the creation of large waves. Permafrost ground is much more sensitive to storm waves" (p. A-14). Since 1934, the coast has been eroding at an average of more than six feet a year. Ten years ago, a huge storm broke off more than forty-two feet of the shoreline. "The joke is," said Calvin Pokiak, assistant land administrator for the Inuvialuit Regional Corporation, "in the houses over here, you have to sleep with your life jacket on" (p. A-14).

The Mackenzie Basin Impact Study (1997) anticipated that the area is warming at three times the global rate, said Kevin Jardine, a Greenpeace climate-change campaigner (Ralston 1996). Other current and future impacts in the same area "include wildlife habitat destruction, melting permafrost which destabilizes northern homes, roads and pipelines, large-scale coastal erosion, and insect outbreaks" (1996). The report suggests that increasing numbers of forest fires in the area may be attributed to rising temperatures.

The Canadian Inuit town of Inuvik, ninety miles south of the Arctic Ocean near the mouth of the Mackenzie River, the temperature rose to 91 degrees F on June 18, 1999, a type of weather unknown to anyone living in the area. "We were down to our T-shirts and hoping for a breeze," said Richard Binder (aged fifty), a local whaler and hunter (Johansen 2001, 19). Along the MacKenzie River, according to Binder, "hillsides have moved even though you've got trees on them. The thaw is going deeper because of the higher temperatures and longer periods of exposure" (p. 19). In some places near Binder's village, the thawing earth has exposed ancestral graves, and remains have been reburied.

Inuit hunters in the northernmost reaches of Canada say that ivory gulls are disappearing, probably as a result of decreasing ice cover that affects the gulls' habitat. Sea ice in the Arctic in the year 2000 covered 15 percent less area than it did in 1978 and has thinned to an average thickness of 1.8 meters, compared with 3.1 meters during the 1950s (Krajick [January 19] 2001, 424). A continuation of this trend could cause the Arctic ice cap to disappear during summers within fifty years, imperiling many forms of life that live in symbiosis with the ice, varying in size from sea algae to polar bears. Less ice means fewer ice-dwelling creatures whose carcasses end up on the ocean floor, feeding clams and other bottom-dwellers, which are, in turn, consumed by walruses. Arctic cod also feed on the ice creatures, grazing the undersides of the ice "like upside-down vacuum cleaners" (p. 424).

THE SPEED OF ARCTIC ICEMELT

In colloquial English, change is said to be "glacial" when one means slow, nearly unchanging. This connotation stems from a long-held belief that climate changes slowly, especially during cold (glacial) periods. Contemporary research, however, indicates that climate changes during glacial periods have often occurred very abruptly. Work by Ganopolski and Rahmstorf (2001) and Hall and Stouffer (2001) indicate that changes in ocean circulation may have been instrumental in these abrupt changes. Temperatures in Greenland are said to have changed by up to 10 degrees C within a matter of decades (Paillard 2001, 147). "The models are not nearly as sensitive as the real world," mentioned Richard B. Alley, an expert on Greenland's climate history at Pennsylvania State University. "That's the kind of thing that makes you nervous" (Revkin [June 8] 2004).

Ice floes in the northern Bering Sea. Courtesy of the National Oceanic and Atmospheric Administration Photo Library.

Late in the summer of 2003, the Ward Hunt Ice Shelf, the largest such formation in the Arctic and quite literally the "mother" of all icebergs in Canada's extreme North, split in half and started breaking up. Lying along the north coast of Ellesmere Island, the 443-square-kilometer ice shelf had supported the Arctic's largest "epishelf" lake—an ice-entombed body of fresh water that provided scientists with a unique natural "cyro-habitat" for some of the Earth's rarest micro-organisms, including algae and plankton from both fresh and salt water. The lake was once hailed as a possible model for explaining how extraterrestrial life might have evolved on Europa, one of Jupiter's frozen moons. The forty-three-meter-deep reservoir of fresh water drained away through the new crack.

The anecdotal observations of the Inuit are being borne out by scientific studies documenting the rapid nature of glacial change in the Arctic. Eric Rignot and Robert H. Thomas, for example, writing in *Science*, described changes in the mass balance of polar ice sheets:

> Perhaps the most important finding of the last 20 years has been the rapidity with which substantial changes can occur on polar ice sheets. As measurements become more precise and more widespread, it is becoming increasingly apparent that change on relatively short time scales is commonplace: Stoppage of huge glaciers, acceleration of others, appreciable thickening, and far more rapid thinning of large sectors of ice sheet, rapid breakup of large areas of ice shelf, and vigorous bottom melting near grounding lines. (2002, 1505)

From their study of Greenland glaciers, H. Jay Zwally and colleagues concluded that warming temperatures accelerate the movement of glaciers, adding to the speed with which they melt. Water that pools on top of glaciers during the summer is conducted to the glaciers' bases, speeding overall movement. In their words, "The indicated coupling between surface melting and ice-sheet flow provides a mechanism for rapid, large-scale dynamic responses of ice sheets to climate warming" (Zwally et al. 2002, 218). The Greenland ice sheet is

Makeshift sea defenses at Shishmaref, Alaska. © Ashley Cooper/Picimpact/Corbis.

losing mass at a rate sufficient to raise sea level by at least 0.13 millimeters per year because of rapid thinning near the coasts (Rignot and
Thomas 2002, 1505).

Laser altimetry measurements published in March 1999 showed rapid
thinning of the eastern portion of the Greenland ice sheet, particularly
in the coastal regions, at rates exceeding a meter a year. The IPCC has
concluded that a 3 degree C warming over Greenland could make
melting of the ice cap irreversible, leading to a seven-meter rise in
global sea level (Leggett 2001, 324).

ACCELERATING DECLINE OF ARCTIC SEA ICE

During 2003, new measurements of sea ice around the North Pole
showed that it had diminished an average of about 4 percent per decade over a twenty-year period. The rate of decline of the perennial ice
(the ice remaining at the end of summer), however, has been nearly

8 percent per decade since 1979 (Stroeve et al. 2005). At the same time, calculations of the age of ice by Fowler, Emery, and Maslanik (2004) indicate that most of the perennial ice pack has shifted from ice older than three years to ice that survives only one or two melt seasons. If this warming trend continues, Arctic sea ice could, within 100 years, disappear almost completely during the summer months, argued Ola Johannessen of the Nansen Environmental and Remote Sensing Center in Bergen, Norway. This change would have a serious impact on wildlife, including the estimated 22,000 polar bears living in the Arctic that need abundant pack ice to hunt for seals during the summer (Connor 2003, 5).

Johannessen led an international team of scientists that measured sea ice across the entire Arctic region using a phenomenon called elastic gravity waves, which utilize vibrations to estimate ice thickness. Given the trends they have detected, this group believes that by 2090 sea ice in the Arctic could diminish by 80 percent from its early century volume (Connor 2003, 5). Johannessen said that reduction of sea ice accelerated significantly after 1980. He believes that the decline of Arctic sea ice is largely due to global warming caused by man-made pollution. "We believe there are strong indications that neither the warming trend nor the decrease of the sea-ice cover over the last two decades can be explained by natural processes alone," Johannessen said (p. 5).

Other researchers have supported Johannessen's assertions. Ice in the Arctic Ocean and on Greenland's ice cap shrank to record-low levels during the summer of 2002, according to Mark Serreze, a polar researcher at the National Snow and Ice Data Center in Boulder, Colorado. "I was really surprised by this," said Serreze. "This was the craziest summer I've seen up there" (McFarling [December 8] 2002). "If we do nothing," said Jonathan Peck, a glaciologist at the University of Arizona, "warming in the next 150 years could be enough to melt the Greenland ice sheet, causing a three-to-six-meter sea level rise that could be catastrophic" ([December 8] 2002).

The dramatic loss of ice in the Arctic, coupled with other new work showing the rise of trees and shrubs across once-barren Arctic tundra lands, "presents a compelling case that something is changing very rapidly over a wide area," said Larry Hinzman, an expert in Arctic change at the University of Alaska (McFarling [December 8] 2002). Weather in the Arctic was unusually warm and stormy during 2001, which broke up ice and melted it more readily (Serreze et al. 2003). "It's the kind of change we'd expect to see," said James Morrison of the University of Washington (Chang [December 8] 2002, A-40).

On Alaska's tundra, where air temperatures have been rising at an average of 0.5 degrees per decade for at least thirty years, shrubs have been observed spreading across land that heretofore has not supported them. A team reporting in *BioScience* during 2005 described processes by which the spread of shrubs tends to reinforce itself: "Increasing shrub abundance leads to deeper snow [shrubs hold more snow than tundra], which promotes higher soil temperatures, greater microbial activity, and more plant-available nitrogen," favoring greater shrub growth in subsequent years (Sturm et al. 2005, 17). "With climate models predicting continued warming," Sturm and colleagues wrote, "large areas of tundra could become converted to shrubland" (p. 17).

Within a century, the melting of the Arctic Ice Cap could lead to summertime ice-free ocean conditions not seen in the area in a million years, the group of scientists wrote during 2005 in *EOS: Transactions of the American Geophysical Union*. Jonathan Overpeck of the University of Arizona and chairman of the National Science Foundation's Arctic System Science Committee and colleagues wrote that the Arctic during the twenty-first century is moving beyond the glacial and interglacial cycle that has characterized the last million years. "At the present rate of change," they wrote, "a summer ice-free Arctic ocean within a century is a real possibility . . . a state [that is] driven largely by feedback-enhanced global climate warming . . . [into] a 'super interglacial' state" (Overpeck et al. 2005, 309).

This report was published days after environmental ministers and officials from twenty-three countries met in Greenland to call on governments to stop arguing over global warming and start acting. That session was held in the town of Ilulissat, near the edge of the Sermeq Kujalleq glacier that has retreated nearly seven miles since 1960 and has become a symbol of fears that the planet is approaching a dangerous warming. "I think probably the biggest surprise of the meeting was that no one could envision any interaction between the components that would act naturally to stop the trajectory to the new system," Overpeck said (Schmid 2005).

ARCTIC ICE NOT DECLINING?

Contrary to the evidence cited by several scientists, not everyone is convinced that Arctic ice is melting significantly. For example, Greg Holloway, a scientist with the Institute of Ocean Science in Victoria, British Columbia, caused a stir at an international meeting of Arctic scientists in Iqaluit, April 25, 2001, by asserting that little evidence exists of a rapid decline in the volume of ice in the Arctic. He suggested instead that the ice has merely been moved around by Arctic winds. "It's more complicated than we thought," said Holloway (Weber 2001, A-8).

Holloway took issue with declassified records compiled by U.S. Navy submarines under the ice as well as satellite pictures that show the surface area of the ice decreasing. The submarine reports indicated that ice volume had diminished 43 percent between 1958 and 1997. According to Holloway, winds blow across the Arctic in a fifty-year cycle. At different points in the cycle, ice tends to cluster in the center of the Arctic. During others, it is blown out to the margins along the Canadian shorelines, where the subs were not allowed because of sovereignty concerns. If the submarines had made their first visit one year earlier and their return one year later, Holloway contended, they would have found no change in the thickness of the sea ice at all (Weber 2001, A-8). Holloway cautioned, however, that his research doesn't suggest a total re-evaluation of the theory of global warming.

Observations of ice in the Canadian Archipelago at the Canadian Ice Center indicate that the ice thickness has actually increased there in recent years. However, other studies have suggested that altered wind fields in the 1990s have acted to "flush" much of the thicker multiyear ice out of the Arctic through Fram Strait, leaving more thin ice (Rigor and Wallace 2004). Calculations of ice age by Fowler,

Emery, and Maslanik (2004) support this hypothesis, as they show that the average age of ice in the Arctic Ocean has declined in recent years.

WARMING AND ALASKA

Glaciers were melting so quickly in Alaska by mid-2005 that their anticipated demise was prompting tourists to visit as quickly as possible. The headline of the Travel Section of the *New York Times* read, "The Race to Alaska before It Melts" (Egan 2005). Cities and towns across the entire state (including Anchorage, Fairbanks, Juneau, and Nome) reported record-high temperatures during the summer of 2004. At Portage Lake, fifty miles south of Anchorage, "people came by the thousands to see Portage Glacier, one of the most accessible of Alaska's frozen attractions. Except, you can no longer see Portage Glacier from the visitor center. It has disappeared" (2005).

Gunter Weller, director of the Center for Global Change and Arctic System Research at the University of Alaska in Fairbanks, contended that mean temperatures in the state have increased by five degrees in the summer and ten degrees in the winter during the last thirty years. Moreover, the Arctic ice field has shrunk by 40–50 percent over the last few decades and lost 10 percent of its thickness. "These are pretty large signals, and they've had an effect on the entire physical environment," Weller said (Murphy 2001, A-1).

Barrow, in northernmost Alaska, experienced a high of 70 degrees F, the highest in memory, during the summer of 2004, Alaska's warmest summer on record, as salmon and porpoises were sighted offshore. Salmon were traditionally so rare off Barrow that indigenous people had to ask their southern relatives how to cure and dry them. Some people swam in Prince William Sound for the first time. Baby walruses were observed swimming in open water, their usual ice-floe homes nowhere to be seen, abandoned by their mothers. Unable to survive on their own, most of the walrus pups died within a month (Petit 2004, 66–69).

Some Alaskan forests have been drowning and turning gray as thawing ground sinks underneath them. Trees and roadside utility poles, destabilized by thawing, lean at crazy angles. The warming has contributed a new phrase to the English language: "the drunken forest" (Johansen 2001, 20). In Barrow, home of Pepe's, the world's north-ernmost Mexican restaurant, mosquitoes, another southern import, have become a problem for the first time. Barrow also experienced its first thunderstorm on record early in the new millennium. Tempera-tures in Barrow began to rise rapidly at about the same time that the first snowmobile arrived, in 1971. By the summer of 2002, bulldozers were pushing sand against the invading sea in Barrow.

For native peoples—17 percent of Alaskans—global warming is a particular threat. The natural world "is our classroom," said Sterling Gologergen, an environmental specialist with the Nome-based Norton Sound Health Corporation (Rosen 2003, 1). In her home region of Alaska, traditional bowhead whaling migrations have been disrupted by rising temperatures. Walrus hunters must travel further, at greater risk, to find animals at the edges of receding ice. Beavers, previously unknown in the region, are showing up in local streams, and their dams could interfere with water quality and fish runs.

In Kotzebue, Alaska, according to a report in the *Guardian* of London, "tundra has turned from spongy to dry" (Campbell 2001, 11). In Fair-banks, portions of some golf courses are collapsing as permafrost begins to melt. North of Fairbanks, houses were lifted on hydraulic jacks to prevent sinking due to melting permafrost, "which is no longer per-manent" (Egan [June 16] 2002, A-1). By 2002, the trans-Alaska oil pipeline was being inspected for damage due to melting permafrost. The pipeline, built during the 1970s, was designed per assumptions that the permafrost would never melt.

Mark Lynas, extracting from his book *High Tide: News from a Warming World* (2004), wrote in the *London Guardian*:

> Roads all around Fairbanks are affected by thawing permafrost; driving over the gentle undulations is like being at sea in a gentle swell. In some places the damage is more dramatic—crash barriers have bent into weird contortions, and wide cracks fracture the dark

tarmac. Permafrost damage now costs a total of $35 million every year, mostly spent on road repairs. Some areas of once-flat land look like bombsites, pockmarked with craters where permafrost ice underneath them has melted and drained away. These uneven landscapes cause "drunken forests" right across Alaska. In one spot near Fairbanks, a long gash had been torn through the tall spruce trees, leaving them toppling over towards each other. ([February 14] 2004, 22)

Across large parts of the Arctic, permafrost has been melted by record-high temperatures during recent decades. "It's really happening almost everywhere," said Vladimir Romanovsky, a geophysicist at the University of Alaska. The accelerating pace of permafrost melting has shocked researchers, according to one observer (Stokstad 2004, 1618). A major reason that researchers are so concerned has been the potential that melting permafrost will release additional carbon dioxide into the atmosphere. For many thousands of years, the Arctic has been locking carbon away in permafrost; estimates range from 350 to 450 gigatons, about a quarter of the Earth's soil carbon. In parts of Siberia, peat deposits that are hundreds of meters thick extend several thousand miles. Permafrost covers about a third of the land in the Northern Hemisphere (p. 1619).

During the last half of the twentieth century, researchers began to detect signs that the Arctic's permafrost stores were re-entering the atmosphere because of rapidly warming temperatures. The area was converting from a "sink" to a net source of greenhouse gases. Walter Oechel, an ecologist at San Diego State University, led research at Toolik Lake and Barrow, Alaska, where they found that the Alaskan tundra was releasing more carbon dioxide than it was absorbing—a surprise that was, according to Oechel, "contrary to all that was known about the Arctic system functioning" (Stokstad 2004, 1620). As the melting of the Arctic's carbon stores continues, the region is "likely to be a huge positive feedback on global warming" (Stokstad, 2004, 1620). The warmer and drier the soil becomes, the more carbon dioxide it releases; warmer, wetter soil releases more methane. As larger volumes of green plants invade the Arctic with warmer weather, however, they may absorb some of the excess carbon dioxide emitted from melting permafrost.

During August of 2005, climate researchers Sergei Kirpotin from Tomsk State University in western Siberia and Judith Marquand from Oxford University returned from the area and said that a large area of western Siberia was undergoing an unprecedented thaw that could dramatically increase the rate of global warming. They said that permafrost across a million square kilometers, an area as large as France and Germany combined, has started to melt (Sample 2005, 1). What was until recently a barren expanse of frozen peat had turned, during the summer, "into a broken landscape of mud and lakes, some more than a kilometer across" (p. 1). In the winter, so much methane is being released that its bubbles have kept the surface from freezing.

Kirpotin told the *Manchester Guardian Weekly* that the situation was an "ecological landslide that is probably irreversible and is undoubtedly connected to climatic warming." He added that the thaw had probably begun in the past three or four years (Sample 2005, 1). "When you start messing around with these natural systems, you can end up in situations where it's unstoppable. There are no brakes you can apply," said David Viner, a senior scientist at the Climatic Research Unit at the University of East Anglia. "This is a big deal because you can't put the permafrost back once it's gone. The causal effect is human activity and it will ramp up temperatures even more than our emissions are doing" (p. 1).

As much as 90 percent of the northern hemisphere's permafrost (the top three meters, or ten feet) could thaw by the end of the twenty-first century given "business-as-usual" increases in greenhouse gas emissions, according to projections by scientists affiliated with the National Center for Atmospheric Research (Lawrence and Slater 2005). Even with major reductions in emissions, Northern Hemisphere the permafrost is projected to shrink from 4 million to about 1.5 million square miles by 2100.

During January and February of 2001, snow was in short supply across much of Alaska. For several weeks, unusual rain soaked areas around Anchorage. The snowless tundra along Norton Sound, near the end of the 1,760-kilometer Iditarod race trail, was so bare in December that parts of it caught fire ("Late Snow" 2001). During training for the race in the Matanuska-Susitna Valley, forty miles north of Anchorage, much of the ground was bare or ice-covered. Some trails were so icy that dogs risked injury. Sufficient snow fell before the race's scheduled start in March, however.

In 2003, for the first time, the Iditarod was postponed for lack of snow. Snow was shipped in for the race's start on March 1, 2003, and the race route was revised to avoid some areas. As in 2001 and 2002, unusual rain soaked parts of the race route, as Alaska experienced its mildest winters in more than a century of recordkeeping. At the same time, the eastern United States was experiencing intense cold and above-average snowfall. Vermont was having its coldest winter in twenty-five years, and many of the big cities on the U.S. Eastern Seaboard experienced repeated heavy snow and ice storms. During 2004, however, Alaska had abundant snow at race time—something that used to be no cause for comment.

The retreat of Alaskan glaciers is most evident via comparison of photographs taken today with those made a century or more ago in the same locations. Bruce Molnia, a geologist with the U.S. Geological Survey, has gathered more than 200 glacier photos taken from the 1890s to the late 1970s. "Where masses of ice were once surging down wide mountain passes into the sea, or were hanging from high and perilously steep faces," wrote David Perlman of the *San Francisco Chronicle*. "What remains from many of the retreating glaciers are stretches of open water or broad, snow-free layers of sediment" (Perlman 2004, A-18). "And as the glaciers disappear," Molnia said, "you get the amazing appearance of vegetation." On the tundra north of Alaska's Brooks Range, according to Molnia's observations (as cited by Perlman), "Explosive bursts of vegetation—willows, alders, birch and many shrubs—are thriving where permafrost once kept the tundra surface frozen in winter. The growth of shrubs across the tundra has increased by 40 percent in less than 60 years, and that perturbation is certainly due to the changing climate" (p. A-18).

"The abundance of Arctic shrubs is . . . increasing, apparently driven by a warming climate," according to an article in *Nature* by Matthew Sturm of the U.S. Army Cold Regions Research and Engineering Laboratory in Fort Wainwright, Alaska, and colleagues (Sturm, Racine, and Tape, 2001, 546–548). "The evidence for increasing shrub abundance is most comprehensive for northern Alaska," Sturm and colleagues wrote. "An extensive comparison of old [1940s] and modern photographs has shown that shrubs there are increasing in size and are colonizing previously shrub-free tundra" (Sturm et al. 2005, 17). Shrubs also are increasing in the western Canadian Arctic, the research team stated, "but there the change is inferred from the recollections of long-time residents. In central Russia, a transect along the Pechore River has shown a decrease in tundra and a corresponding increase in shrubland, but for the vast tundra regions of Siberia, there are currently no data on which to make an assessment." The Sturm group added that, despite the lack of specific data for those areas, "satellite remote sensing studies . . . greatly strengthen the case for pan-Arctic expansion of shrubs" (2005).

GEORGE DIVOKY'S GUILLEMOTS

George Divoky, a researcher affiliated with the University of Alaska's Institute of Arctic Biology, has studied a single colony of Arctic sea birds (guillemots) on a remote barrier island off the northern coast of Alaska, near Barrow, for twenty-seven years. He tracked the egg-laying habits of the birds, which are related to the date that the snow melts. By 2000, the snow was melting almost a month later than it did when he began his work in the 1970s.

Divoky followed the birds during more than ten years of fieldwork before he began to connect their changing egg-laying habits with global warming. During 1995, in response to Vice President Al Gore's task force on climate change, a call went out for data sets: did anyone have information that would shed light on regional climate change? Divoky acquired National Weather Service data describing when snow had melted at Barrow and plotted the dates on a graph. Then he looked at his own data on when the first egg had been laid on Cooper Island and plotted those dates as well. The correlation leapt off the page: from 1975 to 1995, snow was melting in northern Alaska, on average, five days earlier each decade. Over those same twenty years, the date his guillemots laid their eggs was occurring, on average, five days earlier each decade as well.

Guillemots require at least an eighty-day snow-free summer in which to copulate, ovulate, hatch, and fledge their chicks. Because summers averaged less than eighty snow-free days per year in northern Alaska until the 1960s, the birds must have migrated to the area after that decade. Divoky realized that the guillemots were tracking the region's expanding snow-free seasons. He realized that "an earlier date of snow melt was, in effect, an indication that the seasons were in flux; that in a mere 20 years, the brief Arctic summer was now arriving 10 days earlier; and perhaps most important, that climate change was having a biological effect, leaving a fingerprint on a species living in a seemingly remote, pristine environment thousands of miles away from the industrial hand of man" (Frey 2002, 26).

As warmer summers melted the offshore pack ice in its habitat, Divoky's colony was unable to reach the edge of the ice, so they began to die prematurely. Darcy Frey, writing in the *New York Times Sunday Magazine*, described how Divoky began to connect his case study with a worldwide warming pattern:

By 1999, when a series of papers came out describing a major retreat and thinning of the Arctic pack ice due not only to gradually warmer temperatures, but also to a decade-long upper atmospheric shift called the Arctic Oscillation, George was in a position to put the pieces together. His colony was not merely tracking the advancement of snow melt and the earlier arrival of summer; it was also articulating a change in the very makeup of the Arctic itself—the shrinking of the polar ice cap—with all its potentially drastic worldwide consequences. (Frey 2002, 26)

SPRUCE BEETLE OUTBREAKS ON
THE KENAI PENINSULA

On Alaska's Kenai Peninsula, a forest nearly twice the size of Yellowstone National Park has been dying. "Century-old spruce trees stand silvered and cinnamon-colored as they bleed sap" due to spruce bark beetle infestations spurred by rising temperatures, wrote reporter Tim Egan of the *New York Times* ([June 16] 2002, A-1). Temperatures hit the middle eighties there by mid-May 2002, as park rangers worried that tinder-dry, warm conditions could provoke major wildfires. During fifteen years (1988–2003), 40 million spruce trees on the Kenai Peninsula died (Whitfield 2003, 338). The beetle infestations have reached Anchorage, where "visitors flying into the city's airport cross islands covered with the bristling, white skeletons of dead trees that are easily visible through the plane windows" (Lynas [*High Tide*] 2004, 60).

Alaskan author Charles Wohlforth described the coming of the bark beetles:

> On certain spring days in the mid-1990s, clouds of spruce bark beetles took flight among the big spruce trees around Kachemak Bay, 120 miles south of Anchorage. They could be seen from miles away, rolling down the Anchor River valley. People who witnessed the arrival sometimes felt like they were in a horror film, the air thick with beetles landing in their eyes and catching in their hair, and knew when it happened that their trees were destined to turn red and die. (Wohlforth 2004, 238)

The six-legged spruce beetle, which is about a quarter-inch long, takes to the air in the spring, looking for trees on which to feed. When beetles find a vulnerable group of trees, they will signal to other beetles via "a chemical message," Holsten, with the Forest Service in Alaska, said. They then burrow under the bark, feeding on woody capillary tissue that the tree uses to transport nutrients. Healthy spruce trees produce chemicals (terpenes) that usually repel beetles. The chemicals cannot overwhelm a mass infestation of the type that has been taking place, however (Egan [June 25] 2002, F-1). As a spruce dies, green needles turn red, then silver or gray. According to Egan's account, "Ghostly stands of dead, silver-colored spruce—looking like black and white photographs of a forest—can be seen throughout south-central Alaska, particularly on the Kenai. Scientists estimate that 38 million spruce trees have died in Alaska in the current outbreak" (p. F-1).

More than 4 million acres of white spruce trees on the Kenai Peninsula were dead or dying by 2004 from an infestation of beetles, the worst devastation by insects of any forest in North America. Beetles have been gnawing at spruce trees in Alaska for many thousands of years, but with rapid warming since the 1980s, their populations have exploded (Egan [June 25] 2002, F-1). A series of warmer than average years in Alaska has allowed the spruce bark beetles to reproduce at twice their historic rate. "Hungry for the sweet lining beneath the bark," wrote Egan, "the beetles have swarmed over the stands of spruce, overwhelming the trees' normal defense mechanisms.... The dead spruce forest of Alaska may well be one of the world's most visible monuments to climate change." By 2002, nearly 95 percent of the spruce on the Kenai Peninsula had been destroyed by the beetles, leaving a tinder-dry forest ripe for major wildfires that will further devastate the habitats of resident moose, bear, salmon, and other creatures (p. F-1).

"The chief reason why the beetle outbreak has been the largest and the longest is that we have had an unprecedented run of warm summers," said Edward Berg (aged sixty-two), a longtime student of the Kenai Peninsula ecosystem (Egan [June 25] 2002, F-1). Berg has been piecing together the causes of the forest's demise. "His [Berg's] work is very convincing; I would even say unimpeachable," said Glenn Juday, a forest ecologist at the University of Alaska. "For the first time, I now think beetle infestation is related to climate change" (p. F-1). Ed Holsten, who studies insects for the Forest Service in Alaska, said he thinks that climate change is only one reason for the beetle outbreak. The trees on the Kenai are old and ripe for beetle outbreaks, he contends. If they had been logged, or burned, it might have kept the bugs down (p. F-1).

"It's very hard to live among the dead spruce; it's been a real kick in the teeth," said Berg. "We all love this beautiful forest" (Egan [June 25] 2002, F-1).

SHISHMAREF, ALASKA, WASHES INTO THE SEA

Six hundred Alaskan native people in the village of Shishmaref, on the far western shore of Alaska about sixty miles north of Nome, have been watching their village erode into the sea. The permafrost that once reinforced Shishmaref's waterfront is thawing. "We stand on the island's edge and see the remains of houses fallen into the sea," wrote Anton Antonowicz of the *London Daily Mirror*. "They are the homes of poor people. Half-torn rooms with few luxuries. A few photographs, some abandoned cooking pots. Some battered suitcases" (Johansen 2001, 19).

Shishmaref residents voted 161 to 20 during July 2002 to move the entire village inland, a project that the Army Corps of Engineers estimated would cost more than $100 million. Shishmaref is on the Chukchi Sea, which is encroaching steadily as permafrost melts and slumps into the sea. By the summer of 2001, the encroaching sea was threatening rusty fuel tanks holding 80,000 gallons of gasoline and stove oil. "Several years ago," observed Kim Murphy of the *Los Angeles Times*, "the tanks were more than 300 feet from the edge of a seaside bluff. But years of retreating sea ice have sent storm waters pounding, and today just 35 feet of fine sandy bluff stands between the tanks and disaster" (Murphy 2001, A-1). By 2001, seawater was lapping near the town's airport runway, its only long-distance connection to the outside. By that time, three houses had been washed into the sea. Several more were threatened. The town's drinking water supply also had been inundated by the sea. The sea was eight feet from cutting the town's main road and was threatening to wash the town dump out to sea.

In *High Tide* (2004), Mark Lynas described the crumbling of Shishmaref: "We stood under the crumbling cliffs. Robert [Iyatunguk] scuffed the base of it with his boot, and icy sand showered down. Up above us an abandoned house hung precariously over the edge, at least a third of its foundation protruding into thin air. The house next door had toppled over and been reduced to matchwood by the waves" (p. 49).

In the fall of 2004, Shishmaref's beaches retreated still further during vicious storms, which peaked October 18 through 20. The same storms flooded businesses along the waterfront in Nome and damaged power-lines, fuel tanks, and roads in at least a half-dozen other coastal villages.

Because of retreating sea ice, Shishmaref's hunters by 2002 were being forced to travel as far as 200 miles from town for walrus. They also began to use boats to hunt the seals that they used to track over ice. "This year the ice was thinner, and most of the year at least part of the ice was open. We don't normally see open water in December," said Edwin Weyiouanna, an artist who has lived most of his life on the Chukchi Sea (Murphy 2001, A-1). In earlier years, the sea was usually frozen during much of the stormy winter months. With warming, the wind-whipped ocean has eroded Shishmaref's waterfront during many weeks when it once was frozen. The town's residents have come to fear the full moon, with its unusually high tides.

Percy Nayokpuk, a village elder, runs a local store that by 2001 perched dangerously close to the edge of the advancing sea. "When I was a teenager, the beach stretched at least 50 yards further out," said Percy, aged forty-eight (Johansen 2001, 19). As each year passes, the sea's approach seems faster. Five houses have washed into the sea; the U.S. Army has moved or jacked up others. The villagers have been told they will soon have to move. Year by year, the hunting season, which depends upon the arrival of the ice, starts later and ends earlier. "Instead of dog mushing, we have dog slushing," Clifford Weyiouanna (aged fifty-eight) told Antonowicz (p. 19).

The climate has been changing dramatically in Arctic Village, a tiny settlement of about 150 native Alaskans inside the Arctic Circle south of the Brooks Range, about 150 miles from Alaska's Arctic Ocean shore. The rise in wintertime temperatures has been most dramatic. "It used to always be 60 [degrees F] below in the winter, but we don't get that any more," said Kias Peter, a seventy-two-year-old experienced hunter and one of the village elders (Campbell 2001, 11). "We used to have four healthy seasons, but all that is off-balance now," Sarah James, fifty-six in 2001, said. "The treeline has changed; the lakes have dried up. Global warming is very real up here" (p. 11). Climate change in Arctic Village is framed within stories of skinny caribou and hungry grizzly bears, as well as the disappearance of glacier-fed mountain streams and their fish.

ALASKAN GLACIERS SHRINKING

Fewer than twenty of Alaska's several thousand valley glaciers were advancing after the year 2000. Glacial retreat, thinning, stagnation, or a combination of these changes characterize all eleven mountain ranges and three island areas that support glaciers, according to U.S. Geological Survey scientist Bruce Molnia ("Alaskan Glaciers" 2001). "The Earth recently emerged from a global climate event, called the 'Little Ice Age' during which Alaskan glaciers expanded significantly," explained Molnia. The Little Ice Age began to wane in the late nineteenth century. In some areas of Alaska, glacial retreat started during the early eighteenth century, prior to the beginning of the industrial revolution. "During the twentieth century, most Alaskan glaciers receded and, in some areas, disappeared. But it is important to note that our data do not address whether or not any of these changes are human-induced," said Molnia, who warned against blaming the receding glaciers on any single cause, including human emissions of greenhouse gases (2001).

In Glacier Bay, Alaska, 95 percent of the ice observed when the area was first mapped during the 1790s had melted by the year 2000. The first European explorers to visit the area named it after ice that covered nearly the entire harbor. By the 1880s, steady glacial retreat opened liquid water in a forty-mile-long bay. By 2002, the ice that once nearly covered the bay had receded more than sixty miles and was retreating at about half a mile a year ("Alaskan Glaciers" 2001; Toner [June 30] 2002, 4-A).

Some scientists believe that rapid glacial melting in Alaska could be a harbinger of worldwide sea-level rise. University of Colorado professor Mark Meier anticipates that the level of the world's oceans will rise between seven and eleven inches by the end of the twenty-first century, more than twice the level anticipated in 2000 by the IPCC (Russell

McCall Glacier, Alaska, 1958. © Austin Post.

McCall Glacier, Alaska, 2003. © Matt Nolan.

2002, A-4). Meier said that the international panel's previous prediction of a sea-level increase from two to four inches by century's end was too low for several reasons, the most important of which is an underestimation of the water he believes will be contributed by melting glaciers in the Alps, southern Alaska, and the Patagonian mountains of South America (p. A-4).

Recent measurements by aircraft using global positioning satellites and laser altimeters show that the 5,000-square-kilometer Malaspina Glacier in Alaska is losing nearly a meter of thickness per year, the equivalent of three cubic kilometers of water. That glacier alone has ten times the water of all the glaciers in the Alps, which has lost half its ice volume since the 1850s. Whereas climatologists have focused their attention on the great ice sheets of Greenland and Antarctica for signs of melting that could raise sea levels, Meier asserted that the impact of smaller-scale glacial melting has been underestimated (Russell 2002, A-4).

Richard Monastersky, writing in the *New Scientist*, described how the Mendenhall Glacier, behind Juneau, Alaska, is shrinking "before the public's eyes" as 300,000 visitors a year witness its retreat, which accelerated to 100 meters a year in 2000 (Monastersky 2001, 31). Since the 1930s, that glacier has lost about a kilometer of its length. Observations taken at the Juneau airport indicate that average temperatures there have risen 1.6 degrees C since 1943 (p. 32).

Since 1993 Keith Echelmeyer has surveyed the size of ninety glaciers from northern Alaska to the Cascades of Washington. He uses data from satellites and from his own observations, often taken from a single-engine airplane. For reference, Echelmeyer uses U.S. Geological Survey maps created a half-century ago. He has found that 90 percent of the glaciers are losing mass balance, with more summer melting than winter ice accumulation. "The glaciers in Alaska are giving us a clear picture that indeed something is happening to cause them to thin . . . that is climate-related," said Echelmeyer (Monastersky 2001, 31–32). Most of the glaciers are shrinking about a meter a year, but some, such as the Lemon Glacier near Juneau, are losing two to three meters annually. Echelmeyer believes that several of the glaciers he surveys will be gone in 50–100 years.

A study by Anthony Arendt and colleagues at the University of Alaska/Fairbanks used airborne laser altimetry to estimate volume changes of sixty-seven glaciers in Alaska between the mid-1950s and mid-1990s. The profiles they developed were compared with contours on U.S. Geological Survey and Canadian topographic maps made from aerial photographs taken in the 1950s to the early 1970s (Pianin 2002, A-14). According to a report in *Science* (Arendt et al. 2002, 382–386), these glaciers, representing about 20 percent of the glacial area in Alaska and neighboring Canada, have been melting at an average of six feet a year, and some have retreated as much as a few hundred feet annually, a rate that has accelerated during the past seven or eight years.

Arendt and colleagues calculated that Alaskan glaciers are generating nearly twice the annual volume of melting water as the Greenland ice sheet, the largest ice mass in the Northern Hemisphere. According to this study, the Alaskan melt is adding about two-tenths of a millimeter a year to worldwide sea levels. Alaskan icemelt thus accounts for about 9 percent of the sea-level rise during the last century (Arendt et al. 2002, 382).

"The change we are seeing is more rapid than any climate change that has happened in the last 10 to 20 centuries," said Keith A. Echelmeyer, one of the five researchers who prepared the study (Pianin 2002, A-14). The scientists did not speculate whether accelerating melting results from human-induced global warming, natural factors, or a combination. Sallie L. Baliunas of the Harvard-Smithsonian Center for Astrophysics in Cambridge, Massachusetts (a longtime global-warming skeptic), contended that Alaskan glacial melting is due to a dramatic but temporary shift in Pacific Ocean warm water and wind patterns that began in 1976. "It doesn't have the fingerprints of enhanced greenhouse gas concentrations," she contended (p. A-14).

Whatever the cause, "most glaciers have thinned several hundred feet at low elevation in the last 40 years and about 60 feet at higher elevations," said Keith Echelmayer of the University of Alaska at Fairbanks. The ice cover in the Arctic Ocean itself is shrinking by an area the size of the Netherlands each year (Radford 2002, 9).

CLIMATE CHANGE IN SACHS HARBOUR, BANKS ISLAND

Born in 1954, Rosemarie Kuptana grew up in a traditional Inuit hunting society and spoke only Inuvialuktun (the western Arctic dialect of the Inuit language) until the age of eight. Her home community of Sachs Harbour is a Banks Island village of about 120 people on the Beaufort Sea in the Arctic Ocean, about 800 miles northwest of Fairbanks, Alaska. Born in an igloo, Kuptana has been an Inuit weather watcher for much of her life (she was fifty years of age in 2004). Her job was to scan the morning clouds and test the wind's direction to help the hunters decide whether to go out and what everyone should wear.

"We can't read the weather like we used to," said Kuptana. She said that autumn freezes now occur a month later than they did in her youth; spring thaws come earlier, as well. Residents of Sachs Harbour still suffer through winters that most people from lower latitudes would find chilling, with temperatures as low as minus 40 degrees F. Whereas such temperatures once were commonplace during the winter, however, they now are rare. "The permafrost is melting at an alarming rate," said Kuptana (Johansen 2001, 20). Foundations of homes in Sachs Harbour are cracking and shifting due to the melting of permafrost.

Kuptana said that at least three experienced hunters had recently fallen to their deaths through unusually thin ice. Never-before-seen species (including robins, barn swallows, beetles, and sand flies) have appeared on Banks Island. No word exists for robin in Inuktitut, the Inuit language. Growing numbers of Inuit are suffering allergies from white pine pollen that recently reached Banks Island for the first time. At Sachs Harbour, mosquitoes and beetles are now common sights where they were unknown a generation ago. Sea ice is thinner and now drifts far away during the summer, taking with it the seals and polar bears upon which the village's Inuit residents rely for food. Young seals are starving to death because melting and fracturing sea ice separates them from their mothers.

In the winter, sea ice often is thin and broken, making travel dangerous even for the most experienced hunters. In the fall, storms have become more frequent and more violent, making boating difficult. Thunder and lightning have been seen for the first time at Sachs Harbour, arriving with another type of weather that is new to the area—dousing summer rainstorms.

> When I was a child, I never heard thunder or saw lightning, but in the last few years we've had thunder and lightning. . . . The animals really don't know what to do because they've never experienced this kind of phenomenon. We don't know when to travel on the ice and our food sources are getting further and further away. . . . Our way of life is being permanently altered. . . . We have no other sources of food, the people in my community are completely dependent on hunting, trapping and fishing. . . . We have no means of adapting to a different environmental reality, and that is why our situation is so critical. (Johansen 2001, 20)

YUKON CASKETS RIDE MELTING PERMAFROST

Thawing permafrost has been bringing Inuvialuit caskets, many more than eighty years old, to the surface on parts of Herschel Island, Yukon Territory. Animals, including caribou, have tampered with some of the caskets, while others have been looted of artifacts, presumably by people. According to DeNeen L. Brown of the *Washington Post*, "Graves are pushing up from the ground as the ice within the carpet of permafrost melts, churning the soil beneath it into a muddy soup, spitting up foreign contents, sending whole hill slopes sliding downward. On a far tip of this island an entire grave site one day got up and slipped into the sea" (D. L. Brown 2001, A-27). The gravesites are all that remains on the island of a once-thriving Inuvialuit community. The island now is mostly deserted, with only the few park rangers as summer residents. Tourists come by plane to hike, camp, and watch birds.

Brown reported that "the older generation of Inuvialuit believes that anyone who touches the possessions of the dead after they are buried will be cursed. Some in the younger generation, many years removed from traditional life, are wrestling with both sides of the issue. They want to maintain tradition, but they also do not want to sit back while their ancestors' bones lie uncovered on melting ground" (2001, A-27).

Wayne Pollard, professor of geography at McGill University in Montreal, who began studying Herschel Island in 1988, said that the island's landscapes are some of the world's most vulnerable to climate change. "If it were simply coastal erosion and the graves dropping into the ocean and disappearing, that wouldn't be a problem," he said. "Because as I understand it, the Inuvialuit are quite comfortable with the idea of bodies being returned to the environment, as part of the natural cycle that they accept in life" (D. L. Brown 2001, A-27). Most often, however, the graves do not reach the sea.

ICEMELT IN GREENLAND

The largest mass of ice in the Northern Hemisphere resides atop Greenland, which is home to 10 percent of the world's ice. This ice is being measured and monitored as never before, by satellites, by aircraft, and by dozens of down-swaddled scientists who are braving thirty-below-zero temperatures and deadly snow-cloaked crevasses that corrugate the slumping edges of the ice cap (Revkin [June 8] 2004).

Greenland's ice is only a fraction of the size of Antarctica's, but it is melting more rapidly, in part because summers are warmer and thus allowing for more rapid runoff. Philippe Huybrechts, a glaciologist and ice-sheet modeler at the free University of Brussels, has modeled the behavior of Greenland's ice sheet, finding that, with an anticipated annual temperature increase of 8 degrees C, "the ice sheet would shrink to a small glaciated area far inland and sea level would rise by six meters" (Schiermeier 2004, 114). If, over the course of several centuries, global warming provokes the melting of that ice cap, climate models run by several scientists indicate that "removal of the Greenland ice sheet due to a prolonged climatic warming would be irreversible" (Toniazzo, Gregory, and Huybrechts 2004, 21).

Evidence published in *Geophysical Research Letters* late in 2005 (Howat et al. 2005) described a sudden thinning of Greenland's Helheim glacier, on the island's east coast. Professor Slawek Tulaczyk, of the

Department of Earth Sciences at the University of California, Santa Cruz, said that between 2000 and 2005, the glacier retreated four and a half miles. As this glacier has retreated and thinned, effects have spread inland "very fast indeed," said Tulaczyk. "If the 2005 speedup also produces strong thinning, then much of the glacier's main trunk may un-ground, leading to further retreat" (Howat et al. 2005). Because the center of the Greenland ice cap is only 150 miles away, the researchers fear that it, too, may soon be affected (Lean 2005).

The Jakobshavn glacier on Greenland's west coast, which is four miles wide and 1,000 feet thick, has accelerated toward the sea at 113 feet a year. This single glacier alone is reckoned now to be responsible for 3 percent of the annual rise of sea levels worldwide. "We may be very close to the threshold where the Greenland ice cap will melt irreversibly," says Tavi Murray, professor of glaciology at the University of Wales. Professor Tulaczyk added, "The observations that we are seeing now point in that direction" (Lean 2005).

Erosion of Greenland's ice poses practical dangers for indigenous hunters. DeNeen L. Brown of the *Washington Post* described an Inuit hunter's confrontation with glacial ice in Greenland made more dangerous by a warming climate. Aqqaluk Lynge had been chasing a seal. Fear chilled him when the seal dove under the ice and didn't return. Patience when seal hunting is essential, so he waited, but the seal never came back. When an animal begins to act strangely, such as not coming up for breath, something tremendous is happening in nature, wrote Brown.

> The iceberg was at his back. Suddenly it began moving like a monster that was waking up. Lynge . . . looked up in alarm, knowing that these floating mountains, . . . for all their frozen beauty, are ruthless and deadly. So he decided to get mov-ing . . . but the engine on his motorboat wouldn't start. Just then he noticed the iceberg moving. If the tip is moving, he knew, it could mean that one end is moving up and the other end is moving down. . . . A friend in another boat nearby quickly gave him a tow. Soon they were speeding away from the iceberg, not

waiting to look back. Behind them they heard it turning. "We looked back and saw the whole iceberg was collapsing, exploding almost," he said. "We were so afraid." Then it flipped, creating a great tidal wave that crashed hard onto nearby shorelines. By then the two men were out of the wave's path. "When we were finally far away, we could breathe normally again. We were looking back and seeing nothing was left. It exploded underneath the surface of the sea." (2002, A-30)

Although temperatures have been rising rapidly in most of the Arctic, until recently Greenland was not following the trend. Temperatures there now seem to be increasing, however, according to new satellite data analyzed by Dr. Josefino C. Comiso at NASA's Goddard Space Flight Center (Revkin [June 8] 2004). During the last few years, Greenland's "melt zone," where summer warmth turns snow on the edge of the ice cap into slush and ponds of water, has expanded inland, reaching elevations more than a mile high in some places, said Konrad Steffen, a glaciologist at the University of Colorado. Some tongues of floating ice, where glaciers protrude into the sea, are thinning rapidly. Measurements during 2004 by Steffen and others on the Petermann Glacier in northern Greenland indicated that more than 150 feet of thickness had melted away under that tongue in one year. "If other ice streams start to react in a similar way," he said, "then we will actually produce much more fresh water" (2004).

W.S.B. Paterson and Niels Reeh have presented direct measurements of the changes in Greenland's surface elevation between 1954 and 1995 on a traverse of the north Greenland ice sheet. Writing in *Science*, they reported: "We find only small changes in the eastern part of the transect, except for some thickening of the north ice stream. On the west side, however, the thinning rates of the ice sheet are significantly higher, and the thinning extends to higher elevations than had been anticipated from previous studies" (Paterson and Reeh 2001, 60). "The higher elevation appears to be stable, but in a lot of areas around the coast the ice is thinning," said Waleed Abdalati, a manager in the Earth Sciences Department of NASA's Goddard Space Flight

Center. "There is a net loss of ice, particularly in the south" (D. L. Brown 2002, A-30). Whereas some studies suggest that Greenland's ice is melting at increasing rates, one study indicates that temperatures at the summit of the ice sheet have declined at a rate of 2.2 degrees C per decade since 1987 (Chylek, Box, and Lesins 2004, 201). In some places, coastal thinning of ice totaled as much as three feet a year during the 1990s.

Modeling by Jonathan Gregory of the Reading, United Kingdom, Centre for Global Atmospheric Modeling suggests that "as ice is lost, portions of the surface of Greenland's interior will heat . . . at lower elevations where the air is warmer. Less snowfall and more rain would cause the ice to disappear at a faster rate than it is being replaced, leading in turn to further drops in elevation" (Schiermeier 2004, 114–115). What's more, models suggest that the ice sheet would not reappear even if temperatures cooled in the future. The ice sheet creates its own climate, "depending on itself to exist" (pp. 114–115). According to these models, a warming of 3 degrees C could initiate eventual melting of the entire ice sheet over 1,000 years or more, raising global sea levels by about seven meters. Gregory and colleagues have written that "concentrations of greenhouse gases probably will have reached levels before the year 2100 that are sufficient to raise the temperature past this warming threshold" (Gregory, Huybrechts, and Raper 2004, 616). Models are complicated by several factors. For example, warmer temperatures initially could increase snowfall, delaying glacial melting. Icemelt also may depress the Gulf Stream, causing cooling over Greenland.

RUSSIA'S MELTING PERMAFROST AND BURNING PEAT BOGS

Some of the permafrost that covers as much as 65 percent of Russia has been melting; scientists there expect the permafrost boundary to recede 150 miles northward during the next quarter-century. Already, the diamond-producing town of Mirny, in Yakutia, has evacuated a quarter of its population because their houses melted into the previously frozen soil. Parts of the Trans-Siberian Railway's track has twisted and sunk due to the melting of permafrost, causing delays of service of several days at a time.

By 2002, melting permafrost had damaged 300 apartment buildings in the Siberian cities of Norilsk and Yakutsk (Goldman 2002, 1494). "Assuming that the region [Siberia] continues to warm at the modest rate of 0.075 degrees C per year, Lev Khrustalev, a geocryologist at Moscow State University, estimated that by 2030, all five-story structures built between 1950 and 1990 in Yakutsk, a city of 193,000 people, could come crashing down unless steps are taken to strengthen them and preserve the permafrost" (p. 1494).

Russian weather experts described Siberia's problems with melting permafrost during an international conference on climate change in Moscow held in early fall 2003. Georgy Golitsyn, director of Moscow's Institute of Atmospheric Physics, said that, by the end of the twenty-first century, temperature increases in Siberia could be twice the worldwide rise of 1.4–5.8 degrees C anticipated by the IPCC (Meuvret 2003). "Extreme weather events might happen more frequently, [with] the melting of permafrost, which is already noticeable, and damage risks to buildings, roads and pipelines. Pipelines are always having some trouble," he said (2003).

"Here in Moscow and in European Russia, really cold episodes are becoming quite rare. And in Siberia, very heavy frosts have almost disappeared. Instead of minus 40 Celsius or minus 50 which were quite frequent, now these occur just occasionally and they usually experience minus 30," Golitsyn remarked (Meuvret 2003). The disastrous flooding of the Lena River basin in 2001 gave a foretaste of the problems to come, he noted. "The winter had been normal, but . . . the soil was frozen and then in May a heat wave came with temperatures up to 30 degrees [C] when the snow had not melted yet. . . . This is the type of catastrophe we might have more frequently" (2003).

Golitsyn's warning was supported by Michel Petit, who was, until April 2002, a French representative to the IPCC. Siberia, he said, is "one of the most sensitive regions to climate change on the planet" (Meuvret 2003). Although areas that could be farmed would probably expand in Siberia and maritime transport would be opened at seasons when it is now blocked by ice, melting of the permafrost, which covers 60 percent of Russia's surface, would be a "real catastrophe," turning large areas of Siberia into swamps. "Major industrial complexes, towns and pipelines would subside. Carbon dioxide and methane would escape, and the greenhouse effect would become even more serious," said Gorgy Gruza, a Russian climate-change expert who opposes the Kyoto Protocol.

Lakes across the Siberian Arctic are shrinking and drying up, according to a comparison of satellite images taken of 10,000 large lakes over a twenty-five-year period. Scientists found that 125 of the lakes disappeared completely and are now revegetated. Researchers at three U.S. universities described their research on lakes spread across 200,000 square miles of Siberia as they asserted that "Arctic warming has accelerated since the 1980s, driving an array of complex physical and ecological changes in the region" (Smith et al. 2005, 1429). After three decades of rising soil and air temperatures in Siberia, "our analysis reveals a widespread decline in lake abundance and area, despite slight precipitation increases" (p. 1429). Why? Warming temperatures lead to thinning and eventual "breaching" of permafrost near lakes, greatly facilitating their drainage to the subsurface (p. 1429).

Russia also has been beset by burning peat bogs during warmer, drier summers. During September 2002, peat fires around Moscow produced so much smoke that motorists operated with headlights on at noon. The fires, which add carbon dioxide to the atmosphere, were especially intense because of the hot, dry summer.

LAND SPECIES SPREADING NORTHWARD IN THE ARCTIC

Scientific study and indigenous peoples' testimonies strongly support assertions that plant and animal species are moving rapidly northward in the Arctic as the climate warms. Scientists combing archives of aerial photographs have traced the spread of trees and shrubs across previously barren areas of the Alaskan Arctic. Scientists reporting in *Nature* compared new photographs to those taken a half-century earlier. A majority of the sixty-six images showed shrub expansion consistent with warming. "The tree line is definitely moving. You can see the increase," said Ken Tape, a coauthor of the study and a research technician at the University of Alaska in Fairbanks. "The tree line is moving north." The scientists examined areas between the Brooks Range and Alaska's Arctic coast. "In 36 of the 66 repeat photo-pairs, we found distinctive and, in some cases, dramatic increases in the height and diameter of individual shrubs, in-filling of areas that had only scattering of shrubs in 1948–50, and expansion of shrubs into previously shrub-free areas" (Sturm, Racine, and Tape 2001, 546).

Two decades of evidence gathered by satellites indicates that large parts of the Northern Hemisphere have grown greener, probably because of rising temperatures. In a report for *Geophysical Research*, scientists said that vegetation north of the 40th parallel has become measurably greener since 1980. The extra greenness is seen in territories that lie north of New York, Madrid, Ankara, and Beijing. Although the total surface area of vegetation has not changed, the vegetation has apparently increased in density. At the same time, the observations show that the growing season in central Eurasia is now about eighteen days longer than it was twenty years ago. In North America, by the year 2000 it was about twelve days longer than twenty years previously (Cooke 2001, C-3).

"What's interesting is that for the first time we have some indication that this elongation of the growing season—this increase in the amount of green stuff on the planet—is related to temperature," said atmospheric scientist Robert Kaufmann (Cooke 2001, C-3). In the Northern Hemisphere, "we saw that year-to-year changes in growth and duration of the growing season of northern vegetation are tightly linked to year-to-year changes in temperature," said coauthor Liming Zhau (p. C-3). The changes are most evident in the spring and fall, the researchers said. Plants' leaves emerge earlier in spring, and the plants retain their greenery longer in autumn. Spring thus arrives about one week earlier than before, and fall persists for an extra ten days.

POLAR BEARS UNDER PRESSURE

Steady melting of Arctic ice threatens the survival of polar bears, who may meet with their demise before the end of this century. Seymour Laxon, senior lecturer in geophysics at UCL's Centre for Polar Observation and Modeling, said that serious concern exists over the long-term survival hopes for the polar bear as a species. "To put it bluntly," he said, "no ice means no bears" (Elliott 2003; Laxon, Peacock, and Smith 2003, 947). Andrew Derocher, a professor of biology at the University of Alberta, supported Laxon's beliefs. "If the progress of climate change continues without any intervention, then the prognosis for polar bears would ultimately be extinction," he said ("Expert Fears" 2003, C-8). Although reluctant to give an exact date for polar bears' extinction, Derocher said changes in the northern ecosystem are occurring so quickly that the habitat may be incapable of supporting polar bears within 100 years. "What we're looking at is major changes in terms of the distribution of polar bears," he said. "But there are so many

Two polar bears walking across the ice. Courtesy of Corbis.

323

variables to put a number of years on something like extinction of a species. I guess ultimately I'm an optimist that humans will turn the events causing global warming back a bit" (p. C-8). During the summer of 2004, hunters found half a dozen polar bears that had drowned about 200 miles north of Barrow, on Alaska's northern coast. They had tried to swim for shore after the ice had receded 400 miles. A polar bear can swim 100 miles—but not 400 ("Melting Planet" 2005).

The offshore ice-based ecosystem is sustained by upwelling nutrients that feed the plankton, shrimp, and other small organisms, which in turn feed the fish. These then feed the seals, which feed the bears. The native people of the area also occupy a position in this cycle of life. When the ice is not present, the entire cycle collapses.

According to a World Wildlife Fund study, "Polar Bears at Risk," a combination of toxic chemicals and global warming could cause extinction of roughly 22,000 surviving polar bears in the wild within fifty years. Lynn Rosentrater, coauthor of the report and a climate scientist in the WWF's Arctic program, said: "Since the sea ice is melting earlier in the spring, polar bears move to land earlier without having developed as much fat reserves to survive the ice free season. They are skinny bears by the end of summer, which in the worst case can affect their ability to reproduce" ("Thin Polar Bears" 2002).

The same report said that increasing carbon dioxide emissions have caused Arctic temperatures to rise by 5 degrees C during the past century and that the extent of sea ice has decreased by 6 percent in twenty years. By roughly 2050, scientists believe that 60 percent of today's summer sea ice will be gone, which would more than double the summer ice-free season from 60 to 150 days ("Thin Polar Bears" 2002). Lower body weight reduces female bears' ability to lactate, leading to fewer surviving cubs. Already, fewer than 44 percent of cubs now survive the ice-free season, according to the report (2002).

In western Hudson Bay, where warmer temperatures during the 1990s provoked earlier ice melting in the summer and later freezing in the fall, polar bears suffered substantial weight loss. For every week that the ice broke up earlier, bears came ashore twenty to twenty-five pounds lighter, said zoologist Ian Stirling of the Canadian Wildlife Service in

Edmonton (Kerr [August 30] 2002, 1492). Stirling has documented a 15 percent decrease in average weight and number of pups born to polar bears in western Hudson Bay between 1981 and 1998. Such conclusions are amply supported from a number of other sources. According to Michael Goodyear, director of the Churchill (Manitoba) Northern Studies Center, for example, "For every week a bear has not been hunting, it is 10 kilograms (22 pounds) lighter, which can be dangerous as polar bears need to fatten up for the five months in the summer and fall that they are forced to fast" (Clavel 2002). Weight loss affects the bears' ability to reproduce and weakens future generations. Polar bears may soon reach a point where they have lost enough weight to render them infertile. Lara Hansen, chief scientist at the World Wildlife Fund, told the *London Independent*: "If the population stops reproducing, that's the end of it" (Connor 2004).

Ringed and bearded seals, the polar bears' usual prey, are among a number of larger animals that have adapted to life on, in, and under the ice. "Ranging year-round as far as the [North] pole," wrote Kevin Krajick in *Science*, "they never leave the ice pack, keeping breathing holes open all winter, and making lairs under snow mounds. During the spring, the snow lairs camouflage their new pups from polar bears and protect them from cold air" (Krajick [January 19] 2001, 425). Ice in many areas of the Arctic now sometimes melts as early as March, when the seals are having their pups. Because the ice breaks up too early, the pups often have not been fully weaned, so many of them starve or else mature in a weakened state. During the last two decades (1980–2002), ice has been breaking up two weeks earlier than previously. Biologist Lois Harwood of the Canadian Department of Fisheries and Oceans said that ice in the western Arctic broke up three weeks earlier than usual in 1998, dumping many hungry pups into the water before they had been weaned. Adult seals were thinner than usual as well, despite available prey exposed by the early break-up of the ice. "They were starving in the midst of plenty," Harwood said (p. 425). Peary caribou also have been observed falling through unusually thin ice during their migrations.

Without ice, polar bears can become hungry, miserable creatures, especially in unaccustomed warmth. During Iqaluit's weeks of record heat in July 2001, two tourists were hospitalized after they were mauled by a bear in a park south of town. On July 20, a similar confrontation occurred in northern Labrador as a polar bear tried to claw its way into a tent occupied by a group of Dutch tourists. The tourists escaped injury, but the bear was shot to death. "The bears are looking for a cooler place," said Ben Kovic, Nunavut's chief wildlife manager (Johansen 2001, 18).

Until recently, polar bears had their own food sources and usually went about their business without trying to steal food from humans. Beset by late freezes and early thaws, hungry polar bears are coming into contact with people more frequently. In Churchill, Manitoba, polar bears waking from their winter's slumber have found the Hudson Bay ice to be melted earlier than usual. Instead of making their way onto the ice in search of seals, the bears walk along the coast until they get to Churchill, where they block motor traffic and pillage the town dump. Churchill now has a holding tank for wayward polar bears that is larger than its jail for people.

Time magazine described Churchill as follows:

> Polar bears that ordinarily emerge from their summer dens and walk north up Cape Churchill before proceeding directly onto the ice now arrive at their customary departure point and find open water. Unable to move forward, the bears turn left and continue walking right into town, arriving emaciated and hungry. To reduce unscheduled encounters between townspeople and the carnivores, natural-resource officer Wade Roberts and his deputies tranquilize the bears with a dart gun, temporarily house them in a concrete-and-steel bear "jail" and move them 10 miles north. In years with a late freeze—most years since the late 1970s—the number of bears captured in or near town sometimes doubles, to more than 100. (Linden 2000)

Polar bears face other problems related to warming. For example, some polar bear dens have collapsed because of lightning-sparked brush fires on the tundra, which may be increasing because of global warming, according to Stirling, who also commented that the Canadian government should consider fighting fires in prime bear denning areas in the north ("Brush Fires" 2002, 18). Some pregnant bears have been digging dens only to have them collapse, Stirling said. "The fires burn off all of the trees and bushes that are on the upper part of the banks holding the roofs together.... The fires melt the permafrost in the adjacent areas, and in particular the roofs, so there's nothing to hold the roofs together. They just collapse" (p. 18). The bears then must find another denning site where they can give birth over the winter. "The fires are definitely affecting the ability of the bears to use some of the prime areas," said Stirling. He also said that, after a fire, vegetation may require seventy years to grow back to a density suitable for denning sites (p. 18).

Yet another potential risk to polar bears is the increased chance that rain in the late winter will cause polar bear dens to collapse before females and cubs have departed. Scientists surveying polar bear habitat in Manitoba, Canada, have observed large snowbanks used for dens that had collapsed under the weight of wet snow (Stirling and Derocher 1993, 244).

PROBLEMS FOR MUSK OXEN, REINDEER, AND CARIBOU

Retreating ice may imperil musk oxen and caribou populations in the Arctic as well as polar bears. Dwane Wilkin, a reporter for the *Nunatsiaq News*, summarized the consequences of global warming for the Iqaluit area: "The good news is that sailing through the Northwest Passage will finally be a cinch. The bad news? Well, global warming will probably drive musk oxen, polar bears and Peary caribou into extinction. And other species—including humans—will face declining food sources" (Wilkin 1997). This situation could lead to "complete reproductive failure" of the caribou in a worst-case scenario (George 2000).

Arctic precipitation may increase as much as 25 percent with the advent of rapidly rising temperatures within a century. When precipitation falls as heavy, wet snow and ice, caribou and other grazing animals have a harder time finding food during lean winter months. Heavier snow cover may lead to smaller, thinner animals that will be forced to go further afield for food in winter. The same animals will be plagued by increasing numbers of insects during warmer summers. Climate change is already making it tougher for browsing animals in the Arctic, such as caribou and reindeer, to survive the seven months of each year when they must find and eat large amounts of lichen and moss buried beneath snow and ice (Calamai 2002, A-23).

Temperatures near the freezing point are more likely to cause precipitation to fall as freezing rain rather than snow, forming a barrier several centimeters thick. Given warmer weather in the Arctic, chances of ice replacing snow are rising, according to U.S. climate expert Jaakko Putkonen, an Earth sciences professor at the University of Washington. Based on two decades of work at the Spitsbergen Archipelago in the Scandinavian Arctic, Putkonen calculated that five centimeters of rain falling on snow is enough to form an impenetrable ice barrier (Calamai 2002, A-23). "You only need one really big event to have a disastrous effect," said Putkonen. A single heavy rainfall on Russia's snowy Kamchatka peninsula led to the death of 5,000 reindeer, he reported (p. A-23). During the winter of 1996–1997, an estimated 10,000 reindeer died of starvation on Russia's far northeast Chukotka peninsula after ice formed a thick crust over pastures, precluding reindeer grazing. Three thousand Sami in northern Sweden tend reindeer that have been suffering from climate change that has often changed snow to ice. During the 1990s, unusually rainy autumns created ice crust on the moss that feed reindeer. According to one account, "Herd sizes . . . tumbled, bringing not just financial hardship but also the threat of cultural decay" (Roe 2001).

In addition to separating grazing animals from their food, ice also raises temperatures at ground level by sealing in latent heat, encouraging the spread of fungi and toxic molds that attack lichen. Signs of this problem were reported among reindeer in the Scandinavian Arctic; experts say the same threat also exists for more than 3 million caribou in migratory herds that sustain scores of aboriginal communities in northern Canada. Computer modeling suggests that the area vulnerable to such rain-on-snow events could increase by 40 percent worldwide by the 2080s, with significant expansion into parts of central Canada that are home to both woodland and barren ground caribou (Putkonen and Roe 2003, 1188).

"They're the same animal genetically, only those in Europe and Asia have been domesticated while here they roam wild," said Don Russell, a biologist who studies caribou in the Yukon for the Canadian Wildlife Service. "They need about three kilograms dry weight every day— about a garbage-bag full. And they can spend about half their time just digging through the snow with their feet and feeding," said Russell (Calamai 2002, A-23).

In the Arctic National Wildlife Refuge, spring has been arriving earlier; consequently, caribou have had difficulty migrating from wintering areas in time to take advantage of periods of maximum springtime plant growth. The arrivals of springs since 1990 have been the earliest in nearly forty years. By the time the animals reached the plain, their principal food plant had gone to seed. Caribou herd populations could decline significantly should future climate and vegetation patterns prevent proper nourishment of calves. High Arctic Peary caribou and musk oxen may even become extinct.

THREATS TO HARP SEALS IN CANADA

According to an account by Colin Nickerson in the *Boston Globe*, early melting of ice in Canada's Gulf of St. Lawrence "is wreaking havoc on harp seals which give birth on the floes and causing economic hardship for hard-pressed fishermen who depend on the controversial spring hunt" (Nickerson 2002, A-1). Several hundred drowned seal pups washed up on the shores of Newfoundland during 2002 after their mothers, unable to find ice, gave birth in open water. Seal pups need at least twelve days on the ice before they complete nursing.

Fishermen who used to kill the seals with clubs or high-powered rifles in a much-criticized hunt for pelts, vitamin-rich oils, and sex organs for the Asian aphrodisiac trade have found their livelihoods threatened by the retreat of the ice, which is reducing harp seal populations. Seal oil is rich in omega-3 fatty acids that may be helpful in reducing blood cholesterol levels.

In earlier years, ice usually extended from Quebec's Magdalen Islands southward to Prince Edward Island. By late March, according to Nickerson, the floes usually teem with hundreds of thousands of seal mothers and their pups. "In five days of flying over the entire region, we haven't been able to spot a single seal pup," Rick Smith, marine biologist and Canadian director of the anti-sealing group, said from Prince Edward Island. "Usually, there are 200,000 to 300,000 harp seals born in the Gulf of St. Lawrence" (Nickerson 2002, A-1). "The seals need ice, but whether there has been a real reproductive failure this year remains to be seen," said Ian McLaren, professor of biology at Nova Scotia's Dalhousie University and an authority on seals. "One year's loss of pups is not necessarily a catastrophe" (p. A-1).

AN ECONOMIC BOON FROM WARMING: A NORTHWEST AND NORTHEAST PASSAGE

Some industrialists are gleefully anticipating the melting of polar ice so that they can open new shipping lanes and ports, drill for oil, or engage in other profitable activities. Bad news for polar bears, for example, may be good for Pat Broe, a Denver entrepreneur, who bought the derelict port of Churchill, Manitoba, on Hudson's Bay, from the Canadian government in 1997 for about $10 Canadian. By Broe's calculations, once the Arctic ice cap melts, Churchill could bring in as much as $100 million a year as a port on Arctic shipping lanes between Europe, Asia, and the Americas—the fabled Northwest Passage— shorter by thousands of miles than present southerly routes (Krauss et al. 2005). One day during the summer of 2005, residents of Pangnirtung, on the east coast of Baffin Island, were greeted by a surprise: a 400-foot European cruise ship, which had dropped anchor unannounced and sent several hundred tourists ashore in small boats.

Oil companies have been pushing into the frigid Barents Sea seeking undersea oil and gas fields made accessible not only by melting ice but also by advances in technology. But now, as thinning ice stands to simplify construction of drilling rigs, exploration is likely to move even farther north (Krauss et al. 2005). In 2004, scientists found evidence of oil in samples taken from the floor of the Arctic Ocean only 200 miles from the North Pole. All told, one quarter of the world's unexploited oil and gas resources lies in the Arctic, according to the United States Geological Survey (Krauss et al. 2005).

A northern sea route from the Atlantic to the Pacific has been an unrealized dream of mariners and governments in Europe since the sixteenth century. Such a fantasy first arose from a goal to cut nearly 8,000 miles off the trade route between Europe and the newly discovered Spice Islands of the East Indies. Given global warming, a Northwest Passage may soon become commercially viable, along with another route, a "Northeast Passage" along Russia's Siberian coast.

According to a report released during 2002, Gary Brass, director of the U.S. Arctic Research Commission, said that within a decade both the Northwest Passage and the Northern Sea Route could be open to vessels lacking reinforcement against the ice for at least a month in the summer, assuming that recent trends reducing ice coverage continue. By 2050, both routes may be open for the entire summer, according to this report (Kerr [August 30] 2002, 1490).

German explorers asserted on October 10, 2002, that global warming and unusual wind patterns had enabled them to become the first navigators to sail unaided in a yacht through the usually icebound route along Russia's Arctic coast. A twelve-man team, led by the explorer Arved Fuchs (aged forty-nine), made the trip on its fourth attempt west-to-east during the summer in a sixty-foot, sail-driven, wooden trawler that they had converted into a yacht. This Northeast Passage was first plotted in 1879 by the Swedish explorer Adolf Nordenskjold, who negotiated the route along the north Siberian coast in a 370-ton ship driven by sail and steam. His success followed centuries of abortive, sometimes fatal attempts by British and Dutch mariners ("Global Warming and Freak Winds" 2002, 14). The voyage of 8,000 nautical miles in the Fuch expedition's boat, the *Dagmar Aaen*, started from Hamburg during May 2002 and ended in the Russian harbor of Providenya on the Bering Sea. The crew survived on astronaut rations and used satellite pictures of Arctic ice pack movements, as well as a microlight seaplane, to help them get through. Fuchs said the expedition was able to reach Wrangel Island off the northeast coast of Siberia, which is usually icebound all year (p. 14).

ARTIFICIAL HOCKEY ICE IN NUNAVUT

By the winter of 2002–2003, a warming trend was forcing hockey players in Canada's far north to seek rinks with artificial ice. Canada's *Financial Post* reported that "officials in the Arctic say global warming has cut hockey season in half in the past two decades and may hinder the future of development of northern hockey stars such as Jordin Tootoo" ("Ice a Scarce Commodity" 2003). According to the *Financial Post* report, hockey rinks in northern communities were raising funds for the installation of cooling plants to create artificial ice because of the reduced length of time during which natural ice was available. In Rankin Inlet, Tootoo's hometown on Hudson Bay in Nunavut, the community of 2,400 residents installed artificial ice during the summer of 2003 after eight years of lobbying for the funds (2003).

Hockey season on natural ice, which ran from September until May in the 1970s, often now begins around Christmas and ends in March, according to Jim MacDonald, president of Rankin Inlet Minor Hockey. "It's giving us about three months of hockey. Once we finally get going, it's time to stop. At the beginning of our season, we're playing teams that have already been on the ice for two or three months," MacDonald said ("Ice a Scarce Commodity" 2003). According to Tom Thompson, president of Hockey Nunavut, "there are about two-dozen natural ice rinks in tiny communities throughout the territory, but only two with artificial ice." Both are in the capital, Iqaluit.

"In my lifetime I will not be surprised if we see a year where Hudson Bay doesn't freeze over completely," said Jay Anderson of Environment Canada. "It's very dramatic. Yesterday [January 6, 2003], an alert was broadcast over the Rankin Inlet radio station warning that ice on rivers around the town is unsafe. The temperature hovered around minus 12 [degrees] C. It's usually minus 37 there at this time of year" ("Ice a Scarce Commodity" 2003).

GLOBAL WARMING AS A HUMAN RIGHTS ISSUE
IN THE ARCTIC

The Inuit have assembled a human rights case against the United States (specifically the George W. Bush administration) because global warming is threatening their way of life. They have invited the Washington-based Inter-American Commission on Human Rights (IACHR) to visit the Arctic to witness the devastation being caused by global warming. Sheila Watt-Cloutier, president of the Inuit Circumpolar Conference (ICC), which represents 155,000 people inside the Arctic Circle, remarked: "We want to show that we are not powerless victims. These are drastic times for our people and require drastic measures" (P. Brown 2003, 14). Although the IACHR is not a tribunal that can issue binding verdicts, a finding in favor of the ICC could become useful evidence in future attempts to sue the United States on climate-change grounds in international legal forums. Such a finding might also be cited in future suits against corporations in U.S. federal courts if and when legal protection is offered to the atmosphere and other resources held in common.

As Watt-Cloutier explained:

People worry about the polar bear becoming extinct by 2070 because there will be no ice from which they can hunt seals.... But the Inuit face extinction for the same reason and at the same time.... This is a David and Goliath story. Most people have lost contact with the natural world. They even think global warming has benefits, like wearing a T-shirt in November, but we know the planet is melting and with it our vibrant culture, our way of life. We are an endangered species, too.... The ocean is too warm.... Our elders, who instruct the young on the ways of the winter and what to expect, are at a loss. Last Christmas after the ice had formed the temperature rose to 4C [39 F] and it rained. We'd never known it before. (P. Brown 2003, 14)

By the end of 2005, scientific projections were calling for even severe warming-provoked changes in the Arctic. Results of a climate modeling study published in the *Journal of Climate* (Bala et al. 2005, 4531–4544) indicated that "We can either address it [global warming] now, before we severely and irreversibly damage our climate, or we can wait until irreversible damage manifests itself strongly," said Kenneth Caldeira, an author of the study and a climate expert at the Carnegie Institution's Department of Global Ecology, based at Stanford University. "If all we do is try to adapt, things will get worse and worse." The paper's lead author, Govindasamy Bala of the Energy Department's Lawrence Livermore National Laboratory, said it might take twenty or thirty years before the scope of the human-caused changes becomes evident, but from then on there is likely to be no debate (Revkin 2005). While most simulations of the potential human impact on climate have been confined to studying the next 100 years or so, this one ran the model from 1870 to 2300.

The study projected a concentration of carbon dioxide that doubles from preindustrial levels in 2070, triples in 2120, and quadruples in 2160, slightly less than present-day rates of increase. The model is notable as the first that assumes consumption of all known reserves of fossil fuels (Bala et al. 2005, 4531). This model anticipates that the Arctic will see the most intense relative warming, with average annual temperatures in many parts of Arctic Russia and northern North America rising more than 25 degrees F around 2100. Antarctica would warm most intensely between 2100 and 2200.

Given the assumptions of this model, the Arctic tundra nearly disappears, declining from about 8 percent to 1.8 percent of the world's land area by the year 2100. Alaska would lose almost all of its evergreen forests and become mainly a temperate region. Land now residing beneath ice would diminish from 13.3 percent to 4.8 percent of the planet's area. Further south, tropical and temperate forests could expand substantially, so that the two forest types could grow on nearly 65 percent of land surfaces instead of 44 percent in the year 2000 (Revkin 2005).

11 MELTING ANTARCTIC ICE

Antarctica has become a land of climatic paradoxes. Ice sheets and shelves have been melting around the edges of the continent more quickly than anticipated, as warming temperatures speed glaciers' movement into the surrounding oceans. The Antarctic Peninsula is among the most rapidly warming areas on Earth, where large ice shelves have been crumbling into the surrounding seas for several years. Temperatures on the West Antarctic Peninsula have risen 8.8 degrees F in winter since 1950 and 4.5 degrees F in summer (Glick 2004, 33). At the same time, sections of Antarctica's interior have experienced a pronounced cooling trend, while most other areas of Earth have warmed.

Are Antarctic ice sheets thickening or thinning? Is sea ice expanding or contracting? Both questions are open to debate. These debates are of great interest because significant melting of land-based Antarctic ice could raise sea levels and inundate the coastal residences of many hundreds of millions of people. Whatever the outcome of these debates, many observations indicate a climate-change—provoked breakdown in the Antarctic food chain, which begins with krill and ends with penguins and whales. These problems may be intensified by human-caused declines in stratospheric ozone levels.

Even without human intervention, sea levels sometimes have changed very rapidly during glacial cycles in the past. P. U. Clark and colleagues investigated sea-level changes roughly 14,200 years ago that resulted in sea-level surges of about forty millimeters a year over 500 years, much more rapid than the one- to two-millimeter sea-level rise of the twentieth century (Clark et al. 2002, 2438; Sabadini 2002, 2376). These "meltwater pulses" probably originated in Antarctica, mostly from ice-sheet disintegration. As this work is being prepared for the press, several reports are indicating that Antarctic glaciers are speeding their movement toward the sea, especially around the fringes of the West Antarctic ice sheet—a harbinger, perhaps, of accelerating sea-level rise worldwide in coming decades.

Global warming can work in mysterious ways, however, full of counter-indications. Witness the fact that sea-level rise may be slowed by increasing snowfall over Antarctica provoked by rising temperatures. The eastern half of Antarctica is gaining weight, more than 45 billion tons a year, according to a new scientific study. Data from satellites bouncing radar signals off the ground show that the surface of eastern Antarctica appears to be slowly growing higher, by about 1.8 centimeters a year, as snow and ice pile up (Chang 2005, A-22). As temperatures rise, so does the amount of moisture in the air, causing snowfall increases in cold areas such as Antarctica. "It's been long predicted by climate models," said Dr. Curt H. Davis, a professor of electrical and computer engineering at the University of Missouri.

Satellite radar altimetry measurements suggest that the East Antarctic ice sheet north of 81.6 degrees south latitude increased in mass by 45 ± 7 billion tons per year from 1992 to 2003. Comparisons with contemporaneous meteorological model snowfall estimates suggest that the gain in mass was associated with increased precipitation. A gain of this magnitude is enough to slow sea-level rise by 0.12 ± 0.02 millimeters per year (Davis et al. 2005).

"This is the first observational evidence" of ice accumulation in this area (Chang 2005, A-22). The accumulation occurring across 2.75 million square miles of eastern Antarctica corresponds to a gain of 45 billion tons of water a year or, equivalently, the removal of the top 0.12 millimeter of the world's oceans. According to Davis, Antarctica "is the only large terrestrial ice body that is likely gaining mass rather than losing it" (p. A-22).

The data for this observation, from two European Space Agency satellites, cover 1992 to 2003, but because the satellites do not pass directly over the South Pole, they did not provide any information for a 1,150-mile-wide circular area around the pole. Assuming that snow was falling there at the same rate as seen in the rest of Antarctica, the total gain in snowfall would correspond to a 0.18-millimeter-a-year drop in sea levels (Chang 2005, A-22).

A "GENTLE" WINTER ON THE ANTARCTIC COAST

Tom Spears, in the *Ottawa Citizen*, described in 2001, "a gentle Antarctic winter [that] is bringing mild breezes along the seashore, weather for pleasant walks in a light jacket, snow that's mushy enough to make snowballs. This is the depth of winter in the southern hemisphere, yet part of Antarctica remains balmy, even above freezing, as shown by many of the scientific research stations that report daily weather" (Spears 2001, A-1). Scientists from the British Antarctic Survey reported in December 2004 that grass was growing on some parts of the Antarctic Peninsula in places that until recently had been covered by ice year-round (Rohter 2005). Great Britain's Rothera Station on the western shore of the Antarctic Peninsula may be the most rapidly warming locale on planet Earth. Average temperatures there rose 20 degrees F during the last quarter of the twentieth century (Bowen 2005, 33).

On August 7, 2001, midwinter in the Southern Hemisphere, temperatures rose to a balmy minus 1 degree C at Argentina's Orcadas base, on an island near the mainland. On the continent itself, the same day brought highs of minus 3 degrees C in the Jubany and Great Wall research stations and minus 4 degrees C at the Russian Bellingshausen station. "A breeze at Bellingshausen dropped the wind chill to minus 13 degrees C., but that's still pretty mild for Antarctica," Spears wrote. "Yet parts of the Antarctic continue to bask in relative warmth. Tomorrow's forecast is for a high of 5 degrees C. at Palmer Station, and in recent years it has risen as high as 9 degrees C. in winter" (Spears 2001, A-1).

The Antarctic Peninsula is usually the continent's warmest region, with average temperatures of about minus 10 degrees C in midwinter. In recent years, however, the midwinter average has risen to minus 3–4 degrees C, according to climatologist Henry Hengeveld, Environment Canada's global-warming expert (Spears 2001, A-1). Warming on the Antarctic Peninsula has been much more rapid than over the rest of Antarctica; climate proxies indicate that the warming is "unprecedented over the last two millennia" (Vaughan et al. 2003, 243).

Furthermore, measurements of 244 marine glacier fronts on the Antarctic Peninsula that have been draining into the sea (with records dating to 1945) indicated that 87 percent of them are retreating and that "a clear boundary between mean advance and retreat has migrated progressively southward" (Cook et al. 2005, 544). The same study indicates, according to the authors, that "increased drainage of the Antarctic Peninsula is more widespread than previously thought" (p. 544).

REDUCTIONS IN ANTARCTIC SEA ICE AFTER 1950

Wendy Qualye and colleagues examined climate changes on Signy Island, at the confluence of the ice-bound Weddell Sea and the Scotia Sea, bordering Antarctica. They found that "declining permanent ice cover, coinciding with an almost 1 degree C. rise in summer air temperatures has radically affected Signy Island since the 1950s" (Quayle et al. 2002, 645). Photographic estimates suggested that the island's permanent sea-ice cover had receded by about 45 percent since 1951. The open-water period increased by sixty-three days between 1980 and 1993 because of a 0.5 degree C increase in summer air temperature (p. 645).

During 2003, glaciologists from the Australian government's Antarctic Division reported a 20 percent decline in Antarctic sea-ice cover during the previous fifty years. The decline was more pronounced than previously believed; it also contrasted with satellite observations indicating that sea-ice extent may have increased during the past thirty years. Mark Curran, who headed the research team, said that its study of core ice's chemical composition dating to 1840 provided the first long-term record of sustained decline in Antarctic ice. "At first glance, this could appear to be at odds with recent opinion that sea ice is not decreasing and may, in fact, be increasing," he said. "Detection of long-term change is masked by large fluctuations from decade to decade, and it is these decadal fluctuations that have produced apparent short-term increases in the satellite data," Curran continued ("New Research" 2003; Curran et al. 2003, 1203). Curran and colleagues used an ice core taken near Law Dome, in the eastern Antarctic; their findings indicate that "sea ice in that part of the ocean was stable from 1840 to 1950, but has decreased sharply since then" (Wolff 2003, 1164).

In this same study, scientists measured levels of methane sulphonic acid, an atmospheric aerosol produced as a result of phytoplankton activity at the surface of ocean waters, in 160 years of core samples dating back to 1840. Phytoplankton activity in the Southern Ocean is closely linked to sea-ice distribution. "What we can say with data obtained from the ice core is that between 1841 and 1950 there was very little change, but there is a marked decline in sea ice distribution since 1950 of around 20 percent," Curran reported ("New Research" 2003).

ICE SHELVES COLLAPSE

Several ice shelves adjacent to the Antarctic Peninsula have collapsed into the ocean in recent years, becoming spectacular poster-images for global-warming advocates. Kevin Krajick, writing in *Science*, remarked that glaciologists in Antarctica "are keeping an eye on an alarming trend: sudden, explosive calving [of icebergs] in parts of Antarctica. The fear is that if this continues, it may hasten the death of glaciers at an unanticipated rate" (Krajick [June 22] 2001, 2245). Ted Scambos, a glacier expert at the National Snow and Ice Data Center, a joint operation of the Commerce Department and the University of Colorado, was quoted as saying that the fracturing is too rapid to be explained by temperature rises alone. He surmised that "summer temperatures are now high enough to form melt pools on the glacial surfaces, which percolate rapidly into small weaknesses to form crevasses. Once a complex of crevasses hits sea level, sea water rushes in, re-freezes, and the mass blows apart" (p. 2245). Scambos then said that such fracturing might spread to the Ross ice sheet (which is much closer to the South Pole than the Antarctic Peninsula) within about fifty years.

The ice shelves of Antarctica lost 3,000 square miles of surface area during 1998 alone. During March 2000, one of the largest icebergs ever observed broke off the Ross Ice Shelf near Roosevelt Island. Designated B-15, its initial 4,250-square-mile (11,007-square-kilometer) area was almost as large as the state of Connecticut. In mid-May 2002, another massive iceberg broke off the Ross Ice Shelf, according to the National Ice Center in Suitland, Maryland. The new iceberg, named C-19 to indicate its location in the western Ross Sea, was the second to break from the Ross Ice Shelf in two weeks. On May 5, researchers spotted a new floating ice mass, named C-18, measuring about forty-one nautical miles long and four nautical miles wide. An iceberg 200 kilometers (120 miles) long, 32 kilometers (20 miles) wide, and about 200 meters (660 feet) thick calved from the Ross Ice Shelf during late October 2002. The iceberg, called C-19, is one of the biggest observed in recent years, said David Vaughan of the British Antarctic Survey ("Monster Iceberg" 2002).

Vaughan said he believed that the emergence of several very large icebergs from the Ross Ice Shelf in so short a time should not be a cause for worry about global warming.

> That's the normal, natural cycle of the ice shelf. There are areas where we've seen ice shelves retreating over the last 50 years, and we think that is a response to climate change, but this is not one of those areas. The Antarctic Peninsula is where climate has been changing most rapidly and where the ice shelves have been retreating. But they don't retreat like this, producing one big iceberg every now and again. They retreat by year-on-year production of lots of little icebergs, a kind of constant retreat. ("Monster Iceberg" 2002)

The Antarctic Peninsula is a different story. Disintegration of ice shelves there seems to be bona fide evidence of rapid climate change.

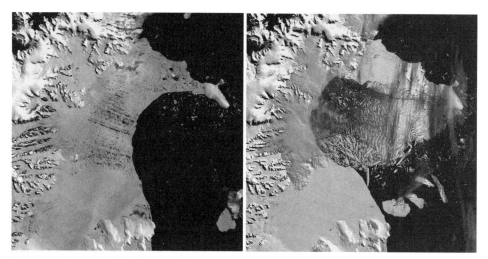

Between January 31 (left photo) and March 5, 2002 (right photo) orbiting satellites observed the collapse of the Larsen B Ice Shelf's northern section. Courtesy of NASA's Terra satellite, supplied by Ted Scambos, National Snow and Ice Data Center.

Near the northernmost tip of the peninsula, the Larsen ice shelves (known by scientists as "A," "B," and "C" from north-south) have been disintegrating since 1995 when the "A" shelf fell apart; "B" followed in 1998, losing an estimated 1,250 square miles over four years. A 1,250-square-mile section of the Larsen B Ice Shelf disintegrated in just thirty-five days, setting thousands of icebergs adrift in the Weddell Sea. "We knew what was left of the Larsen B ice shelf would collapse eventually, but this is staggering," said David Vaughan, a British glaciologist. "It's just broken apart. It fell over like a wall and has broken as if into hundreds of thousands of bricks. . . . This is the largest single event in a series of retreats by ice shelves in the peninsula over the last 30 years" (Vidal [March 20] 2002, 3). The collapse of the Larsen B Ice Shelf "is unprecedented during the Holocene"; that is, during the last 10,000 years, according to a scientific team writing in *Nature* (Domack et al. 2005, 681).

Larsen "C" shattered and collapsed during mid-March 2002. The 640-foot-thick ice shelf had been receding for at least a decade, but scientists said it collapsed with "staggering" speed. The "B" and "C" shelves were believed to have been intact since the last ice age (Toner [March 20] 2002, A-1). Collapse of the Larsen Ice Shelves was attributed to strong climate warming in the region, according to the U.S. government's Ice Center (Vidal [March 20] 2002, 3). "This area and that of the western Arctic off Alaska are the two most rapidly warming places on the globe. The trends of melting ice shelves is now clear," said Steve Sawyer, a climate-change scientist (Vidal [March 20] 2002, 3).

Scambos said that the surprising speed of the ice sheets' collapse, which he blamed on "strong climate warming in the region," will force scientists to reassess the stability of Antarctica's other ice shelves (Toner [March 20] 2002, A-1). In fact, satellite images indicated that during 2002 another massive iceberg, larger in area than Delaware, broke away from the Thwaites ice tongue, a sheet of glacial ice that extends into the Amundsen Sea nearly a thousand miles from the Larsen ice sheet.

The collapse of ice shelves does not, by itself, raise sea levels, but their demise could speed the melting of land-based ice that could accelerate sea-level rises in the future. "Loss of ice shelves surrounding the continent could have a major effect on the rate of ice flow off the continent," said Scambos (Toner [March 20] 2002, A-1). At least five of the ice shelves on the Antarctic Peninsula are now receding faster than at any time in recent history.

After 2002, the breakup of the Larsen Ice Shelf increased glacial movement toward the sea, according to data from two satellites. Some theorists and climate modelers anticipated in the 1970s that global warming would provoke ice shelves in the oceans around Antarctica to melt and release glaciers, increasing the amount of ice being pushed into the sea. "People thought they were wrong, because they thought those models were simplistic," said glaciologist and remote-sensing specialist Eric Rignot of the Jet Propulsion Laboratory in Pasadena, California. "What we are seeing now is that they were not wrong," he said. "They were right" (Nesmith 2004, 9-A). Rignot and Ted Scambos of the National Snow and Ice Data Center at the University of Colorado are the lead authors of two articles published in *Geophysical Research Letters*. Their work indicates a two- to sixfold increase in "centerline speed" for glaciers that were left exposed by collapse of the Larsen B Ice Shelf. One glacier, Hektoria, lost about thirty-eight meters of height within six months (Rignot et al. 2004; Scambos et al. 2004). These observations provide a glimpse of what could happen on a larger scale if other large ice shelves in Antarctica—for example, the huge Ross Ice Shelf—break up. "These glaciers are themselves too small to contribute any significant amount of ice to the sea-level problem," Rignot said, "but the Ross Ice Shelf buttresses much larger glaciers. If these are released, the impact would be much greater" (Nesmith 2004, 9-A).

Some scientists compare the "tongue" of a glacier (where the body of ice meets the sea) to a cork in a bottle. "The tongue of the glacier or the cork in the bottle do not represent that much," said Claudio Teitelboim, director of the Center for Scientific Studies, a private Chilean institution that cooperates with NASA to survey the ice fields of Antarctica and Patagonia. "But once the cork is dislodged, the contents of the bottle flow out, and that can generate tremendous instability" (Rohter 2005).

Glaciers flowing into Antarctica's Amundsen Sea and that help drain the West Antarctic ice sheet were thinning twice as fast near the coast by 2004 as they had in the 1990s. Warmer seawater erodes the bond between coastal ice and the bedrock below, "like weakening the cork in a bottle," said Robert H. Thomas, a glacier expert for NASA in Wallops Island, Virginia. "You start to let stuff out" (Revkin [September 24] 2004, A-24). Thomas and several coauthors wrote in *Science* during 2004 that "recent aircraft and satellite laser altimeter surveys of the Amundsen Sea sector of West Antarctica show that local glaciers are discharging about 250 cubic kilometers of ice per year to the ocean, about 60 per cent more than is accumulated within their catchment basins" (Thomas et al. 2004, 255). At present, such discharge could raise world sea levels about 0.2 millimeters per year—not a startling amount. However, the long-term implications of such ice flow may be more ominous: "Most of these glaciers flow into floating ice shelves over bedrock up to hundreds of meters deeper than previous estimates, providing exit routes for ice from further inland if ice-sheet collapse is underway" (p. 255).

The anticipated thinning of the West Antarctic ice sheet is being intensely examined by scientists. Andrew Shepherd and colleagues, writing in *Geophysical Research Letters*, have used satellite altimetry to associate thinning ice in the area to melting caused by the motion of warming ocean currents (Shepherd, Wingham, and Rignot 2004). In a related study, Anthony J. Payne and colleagues used an ice-flow model to describe how thinning of the Pine Island Glacier affects the ice sheet further inland (Payne et al. 2004).

THE VELOCITY OF ICEMELT: A SLOW-MOTION DISASTER?

Evidence from Antarctica suggests that melting ice may flow into the sea much more easily than earlier believed, perhaps leading to an accelerating rise in sea levels. A study published March 7, 2003, in the journal *Science* suggested that seas might rise as much as several meters during the next several centuries, given projected global warming based on business-as-usual usage of fossil fuels. The study called that prospect "a slow-motion disaster," the cost of which—in lost shorelines, salt inundation of water supplies, and damaged ecosystems—"would be borne by many future generations" (de Angelis and Skvarca 2003, 1560; Revkin 2003, A-8).

The analysis focused on the disintegration of ice shelves at the edges of the Antarctic Peninsula following decades of warming temperatures. The loss of the coastal shelves caused a drastic speed-up in the seaward flow of inland glaciers. The peninsula, which stretches north toward South America, has warmed an average of 4.5 degrees over the last sixty years, so much that ponds of melted water now form during summer months atop the flat ice shelves (Revkin 2003, A-8).

Two Argentine researchers described aerial surveys they conducted during 2001 and 2002 that indicated that the collapse of the Larsen A Ice Shelf during 1995 led to a sudden surge in the seaward flow of five of the six glaciers—as if a doorstop had been removed or a dam breached. Geological evidence indicated no signs of similar ice breakups along the peninsula in many thousands of years, the researchers and other experts said. Indeed, the recent disintegration of ice shelves along both coasts of the peninsula occurred after thousands of years of relative stability, according to Pedro Skvarca, an author of the study and the director of glaciology at the Antarctic Institute of Argentina (Revkin 2003, A-8).

"We are witnessing a very significant warning sign of climate warming," Skvarca said (Revkin 2003, A-8). "This discovery calls for a reconsideration of former hypotheses about the stabilizing role of ice shelves (de Angelis and Skvarca, 2003, 1560). "It should be emphasized that the grounded ice on the northeastern Antarctic Peninsula is rapidly retreating and therefore substantially contributing to the global rise in sea level. The risk increases when the possible surging response of the Kektoria-Green-Evans and Crane glaciers is considered; these glaciers formerly nourished the section of the Larsen B Ice Shelf that disintegrated in early 2002" (p. 1562).

Andrew Revkin reported in the *New York Times* that "the sliding could be abetted not only by the loss of the ice-shelf blockade, they said, but also by another unpredicted result of warming noted by other scientists in Antarctica and in Greenland: the rapid percolation of water from summertime ponds high on the ice sheets down through cracks to the base. There the water acts as a lubricant, facilitating the slide of glacial ice over the earth below" (2003, A-8).

Many of Antarctica's ice shelves may behave in ways similar to those that were studied. This similarity gives the findings great significance, said Scambos. The probability and timing of such an outcome remains unknown, but the new work has shed some light on the question and has increased some scientists' feelings of urgency (Revkin 2003, A-8).

ICE SHEET THINNING IN ANTARCTICA

Speculation regarding whether retreat of West Antarctic glaciers could accelerate ice flow from the continent's interior has engaged glaciologists for a quarter-century. The stakes of this debate are enormous for the many hundreds of millions of people who live in urban areas on and near the world's coasts. Disintegration of the West Antarctic ice sheet could raise world sea levels by roughly six meters, or twenty feet.

Researchers from University College, London, and the British Antarctic Survey reported in *Science* that the Pine Island Glacier, the largest on the West Antarctic ice sheet and an important influence on the movement of the entire West Antarctic ice sheet, has lost thirty-two cubic kilometers of ice over a 5,000-square-kilometer area since 1992. The glacier was losing between one and two meters of thickness per year during the 1990s, according to this study. If thinning continued at such a rate, the entire glacier could disappear into the ocean within a few hundred years. "It is possible," wrote Andrew Shepherd and colleagues, "that a retreat of the Pine Island Glacier may accelerate ice discharge from the West Antarctic Ice Sheet interior.... The grounded PIG thinned by up to 1.6 meters per year between 1992 and 1999, affecting 150 kilometers of the inland glacier" (Shepherd et al. 2001, 862). "If sufficiently prolonged, the present thinning could affect the flow of what is now slow-moving ice in the interior, increasing the volume of rapidly drained ice" (p. 864). During November 2001, scientists discovered a fifteen-mile crack in the Pine Island Glacier, which split off a 150-square-mile iceberg (Toner [March 20] 2002, A-1). This break indicated that the glacier may be sliding toward the ocean more quickly than previously thought.

Andrew Shepherd, who led the study, said: "We have shown for the first time that such a retreat is indeed occurring. It is of paramount importance to determine whether the thinning is accelerating. Our present theoretical understanding is not sufficient to predict firmly the future evolution of the Pine Island glacier" (Radford 2001, 9; Shepherd et al. 2001). This study added weight to the argument that small changes at the coast of the continent provoked by global warming could be transmitted swiftly inland, leading to an acceleration of glacial melting and, ultimately, significant worldwide sea-level rise.

Eric Rignot and Stanley S. Jacobs described the process in *Science*:

As continental ice from Antarctica reaches the grounding line and begins to float, its underside melts into the ocean. Results obtained from satellite radar interferometry reveal that bottom melt rates experienced by large outlet glaciers near their grounding lines are far higher than generally assumed. The melting rate is positively correlated with thermal forcing, increasing by 1 meter per year for each 0.1 degree C. rise in ocean temperature. Where deep water has direct access to grounding lines, glaciers and ice shelves are vulnerable to ongoing increases in ocean temperature. (2002, 2020)

Shepherd and colleagues have also explored the dynamics relating to the disintegration of the Larsen ice shelves on the Antarctic Peninsula. While air temperatures have been rising rapidly in the area, they have not risen quickly enough to wholly explain the rapid collapse of the ice shelves. The researchers believe that melting is being accelerated not only by rising air temperatures above the ice shelves but also by warming water temperatures below (Shepherd et al. 2003, 856). "If so," wrote Jocelyn Kaiser in *Science*, "the rest of the Larsen [Ice Shelf] is doomed, and other Antarctic ice shelves could be more endangered than has been thought" (Kaiser 2003, 759). This one-two punch of warming from atmosphere and ocean during summertime causes pools of melted water to force their way through the ice shelves, producing crevasses that lead to collapse. Although some glaciologists caution that evidence of ocean warming is scanty, others believe that the theories of Shepherd and colleagues help explain why some ice masses are collapsing so quickly (p. 759).

ARE WEST ANTARCTIC GLACIERS THICKENING?

Some studies seem to contradict others. Is the West Antarctic ice sheet increasing its velocity toward the ocean, thinning along the way, or is the opposite occurring? Some of each may be happening at the same time, in different parts of the ice cap. One study (Joughin and Tulaczyk 2002, 476–480) argued that the flow of these great rivers of ice is slowing, and as a result, they are growing thicker. According to Andrew Revkin, writing in the *New York Times*:

The change means that this part of western Antarctica is likely to serve as a frozen bank for water instead of a source, slightly countering an overall trend toward rising seas, according to the research. . . . [The change] appears to be related to the slow warming that has been going on since the end of the last ice age, 12,000 years ago, and not the accelerated warming trend in the last five decades that many scientists have ascribed in part to human activities, according to authors Ian R. Joughin, an engineer at the Jet Propulsion Laboratory of the National Aeronautics and Space Administration in Pasadena, Calif., and Slawek Tulaczyk, a professor of earth sciences at the University of California, Santa Cruz. ([January 18] 2002, A-17)

Joughin and Tulaczyk presented evidence that parts of the ice sheet are growing thicker:

We have used ice-flow velocity measurements from synthetic aperture radar to reassess the mass balance of the Ross Ice Streams, West Antarctica. We find strong evidence for ice-sheet growth (+26.8 gigatons per year) in contrast to earlier estimates (−20.9 gt/yr). . . . The overall positive mass balance may signal an end to the Holocene retreat of these ice streams. . . . [Though there is] ample evidence for a large retreat of the West Antarctic ice sheet over the last several thousand years, . . . if the current positive imbalance is not merely part of decadal or century-scale fluctuations, it represents a reversal of the long-term Holocene retreat. (2002, 476, 479)

Joughlin and Tulaczyk acknowledged that other parts of the West Antarctic ice sheet are behaving in the opposite way, speeding up and sending more ice toward the sea where great icebergs split from the broad ice shelf and eventually melt. "This teaches us more about the system," said glaciologist Richard B. Alley. "But it shouldn't affect what a coastal property owner thinks one way or the other" (Revkin [January 18] 2002, A-17).

In their study, Joughin and Tulaczyk said that the slowing and eventual thickening of the ice might result from thinning during thousands of years since the last ice age. As the rivers of ice lose mass, extreme cold at the surface may more easily migrate through the ice and freeze any water acting as a lubricant deep beneath, causing the ice to stick to the earth and grind to a halt. "In the long run, this could produce a stop-and-start rhythm," said Robert A. Bindschadler, a glaciologist at the NASA Goddard Space Flight Center in Greenbelt, Maryland. "Once the ice streams start to thicken, the chill will not reach the bottom as easily, and the geothermal heat from below might melt some ice, allowing the glacier to start sliding toward the sea again" (Revkin [January 18] 2002, A-17). Whatever is happening in the streams feeding the Ross Ice Shelf is not happening in warmer areas further north, where the Pine Island and Thwaites glaciers are calving more icebergs than ever, said glaciologist Rignot (p. A-17).

Alley cautioned that the record is too short to trace long-term trends from this study: "It is important to remember how short the instrumental record is and how poorly characterized the natural variability. Sedimentary records indicate that ice streams have paused or even re-advanced during the retreat since the last ice age" (Alley 2002, 452). Factors other than global warming (such as basal melting stemming from heat that leaks from the Earth's interior) also may influence the advance or retreat of ice sheets over short periods. Alley suggested that the Pine Island Bay drainage "is probably the most likely . . . to experience the onset of dramatic ice-sheet changes. Here, thick, fast-moving ice discharges into relatively warm ocean waters without the protection of a large shelf" (p. 452).

Regardless of the thickening of some ice sheets, marked decreases in salinity measured in the Ross Sea during the last four decades indicate that the West Antarctic ice sheet is melting. According to S. S. Jacobs and colleagues writing in *Science*, "These changes have been accompanied by atmospheric warming on Ross Island, ocean warming at depths of more than 300 meters north of the continental shelf, . . . and thinning of the southeast Pacific ice shelves" (2002, 386). The freshening of water in the Ross Sea (its decline in salinity), according to this study, "appears to have resulted from a combination of factors, including increased precipitation, reduced sea-ice production, and increased melting of the West Antarctic Ice Sheet" (p. 386).

IS SOME ANTARCTIC SEA ICE EXPANDING?

According to a report by the Environment News Service during 2002, satellite records of sea ice around Antarctica indicate that Southern Hemisphere ice cover has increased since the late 1970s, at the same time that Arctic sea ice has declined. Continued decreases or increases could have substantial impacts on polar climates, because sea ice spreading over large areas increases albedo, reflecting solar radiation away from the Earth's surface ("Antarctic Sea Ice" 2002).

Claire Parkinson of the Goddard Space Flight Center analyzed the length of the sea-ice season throughout the Southern Ocean to obtain trends in sea-ice coverage. Parkinson examined twenty-one years (1979–1999) of Antarctic sea-ice satellite records and found that, on average, the area where southern sea-ice seasons have lengthened by at least one day per year is about twice as large as the area where sea-ice seasons have shortened by at least one day per year. One day per year equals three weeks over the twenty-one-year period. Parkinson also reported that the area of sea ice in the Arctic was decreasing by 13,000 square miles a year—an icepack the size of Maryland and Delaware (Russell 2002, A-4).

ANTARCTIC WARMING THREATENS THE FOOD WEB FROM PHYTOPLANKTON TO PENGUINS

By late 2004, research reported in *Nature* indicated that the amount of krill in the Antarctic marine food chain had fallen by 80 percent since the 1970s, creating food shortages that were endangering larger animals and birds such as whales, seals, penguins, and albatrosses, especially in the vicinity of the Antarctic Peninsula. Angus Atkinson of the British Antarctic Survey, who led the new research, commented: "This is the first time that we have understood the full scale of this decline. Krill feed on the algae found under the surface of the sea-ice, which acts as a kind of nursery" (Atkinson et al. 2004, 100–103; Henderson 2004).

The collapse of ice shelves along some of Antarctica's shores changes the ecology of the nearby ocean, with attendant effects on wildlife. According to a report by the Environment News Service:

> The new icebergs have changed the Antarctic ecosystem, scientists say, blocking sunlight needed for growth of the microscopic plants called phytoplankton that form the underpinning of the entire food web. They are a primary food source for miniscule shrimp-like krill, which in turn are consumed by fish, seals, whales and penguins. Ice shelf B-15 broke into smaller pieces that prevented the usual movement of sea ice out of the region, said Kevin Arrigo, assistant professor of geophysics at Stanford University. Phytoplankton require open water and sunlight to reproduce, so higher-than-usual amounts of pack ice cause declines in plankton productivity. ("Breakaway Bergs" 2002)

Adelie penguins at a rookery. Courtesy of the National Oceanic and Atmospheric Administration Photo Library.

Populations of Adelie penguins on the Antarctic Peninsula are decreasing as their surroundings warm. Around 1985, the Biscoe region of the Antarctic Peninsula was home to about 2,800 breeding pairs of Adelie penguins. By the year 2000, that number had declined to about 1,000. On nearby islands, the number of breeding pairs has dropped from 32,000 to 11,000 in thirty years. In some cases, melting ice has provided an evaporative surface that increases snowfall, which inhibits the birds' nesting. "The Adelies are the canaries in the coal mine of climate change in the Antarctic," said ecologist Bill Fraser (Montaigne 2004, 36, 39, 47). If warming continues, penguins may abandon much of their 900-mile-long promontory home altogether. The archetypal "tuxedoed" species prefer a cold climate even more so than do other penguins (Lean 2002, 9).

Warming has also caused problems for penguins in the Ross Sea. Large icebergs have blocked the way between their breeding colonies and feeding areas. As a result, the penguins are being forced to walk an extra thirty miles (at a one-mile-per-hour waddle) to get food. Thousands of penguins have died during these treks. Thousands of emperor penguin chicks drowned near Britain's Halley base after ice broke up earlier than usual, before they had learned to swim (Lean 2002, 9). The penguins cannot fly and so have trouble changing their habitat as conditions evolve.

According to a report by Geoffrey Lean in the *London Independent*:

[The penguins] are feeling the heat most strongly on the Antarctic Peninsula, which juts out from the polar landmass towards South America. Studies of air temperatures around the world over the past half-century suggest that this is one of the three areas on the planet—along with north-western North America and part of Siberia—warming up fastest. The British Antarctic Survey says flowering plants have spread rapidly in the area, glaciers are retreating, and seven huge ice sheets have melted away. (2002, 9)

Professor Steven Emslie of the University of North Carolina believes that if the warming continues the penguins will "continue to decline in the peninsula, and may completely abandon much of it" (Lean 2002, 9). Studies he has conducted of fossilized remains near Britain's Rothera base show that the number of penguins has declined sharply during warmer periods in the past. Researchers for the U.S. National Science Foundation said that one colony of Adelies at Cape Royds will "fail totally." Scientists at the Scripps Institute of Oceanography added that a colony of emperor penguins at Cape Crozier also has failed to raise any chicks (p. 9).

The Ross Sea is home to 25 percent of the world's emperor penguins and 30 percent of Adelie penguins. "We know for certain that penguins suffered breeding losses because of the icebergs in this region," said Assistant Professor Arrigo of Stanford University.

> There was a lot less food nearby for penguins to get to, so they had to go much farther to feed. In doing so, they left their nests exposed for longer periods of time than they normally would. That made them vulnerable to predators such as the skua, a large gull that feeds on chicks and the eggs. So penguin breeding success was much lower last year [2001]. . . . Now the penguins have another obstacle they have to get around. Not only do they have to go farther to find food, but they have to swim around this enormously large iceberg that has found its way in their path. Some rookeries have been abandoned altogether. ("Breakaway Bergs" 2002)

The increasing difficulty of the penguins' lives has cut their reproduction rates. Penguin chick numbers have fallen to about 10 percent of the usual number at Cape Bird and 2 percent of the usual number at Cape Royds and Cape Crozier. At the southernmost colony of Adelie penguins at Cape Royds, only about 500 nests were established in areas where penguins usually maintain 3,000 to 3,500 active nests, imperiling the entire colony's survival. By 2003, the penguins required four days to complete foraging trips for food, sapping time and energy used previously for reproduction. The U.S. Coast Guard icebreaker *Polar Sea*, which supplies the McMurdo base, broke a path through nearby ice to improve the penguins' access to the sea (English 2003).

Climate change may be a casual factor in the declining numbers of penguins, seals, and albatrosses near the island of South Georgia in the western Antarctic. Melting Antarctic ice may be eliminating krill, one of the oceans' major food sources. British scientists have said that krill is finding it harder to graze on algae beneath retreating ice shelves, making life more difficult for the animals that feed on krill (Reid and Croxall 2001). Krill populations are declining in the Northern Hemisphere as well. They fell 70 percent during ten years in the estuary and gulf of the St. Lawrence, according to new research by scientists with the Maurice Lamontagne Institute, a marine science center associated with Canada's Department of Fisheries and Oceans. A probable cause, the scientists said, is global warming, with a further risk of a reduction in the number of whales and fish in these waters.

Keith Reid and John Croxall of the British Antarctic Survey in Cambridge surveyed the numbers and breeding success of Antarctic fur seals (*Arctocephalus gazella*), macaroni penguins (*Eudyptes chrysolophus*), gentoo penguins (*Pygoscelis papua*), and black-browed albatrosses (*Thalassarche melanophrys*). Since the 1980s, the researchers found that the link between these species and their favorite prey, Antarctic krill (*Euphausia superba*), has changed. "Krill graze on algae beneath sea ice," Reid explained, so retreating sea ice in the region may have reduced the krill's ability to reproduce (Loeb 1997). The amount of krill in the diet of seals and penguins closely matches their ability to bear healthy offspring.

Writing in *Science*, however, D. G. Ainley and colleagues took issue with assertions by Croxall and others that populations of Adelie penguins are declining along the Antarctic Coast because of sea-ice retreat. This is true "only along the Northwestern Antarctic Peninsula (about 5 per cent of the Antarctic Coast) in the last 50 years," they asserted. Elsewhere, "this species has been increasing" (Ainley et al. 2003, 429). Ainley and colleagues further asserted that the penguins depend on krill only in the summer.

Late in 2000 and early in 2001, during their summer breeding season, thousands of Magellanic penguins washed up dead on the beaches north of Punta Tombo, in the far south of Argentina. According to a report in the *New York Times*, "Many birds abandoned their nests, leaving chicks to starve. Among the survivors, many were in bad shape, having difficulty finding the fish they needed to sustain themselves" (Yoon 2001, F-1). Since 1987, the number of Magellanic penguins at Punta Tombo has declined by 30 percent (p. F-1).

Penguin populations are declining around the world, and global warming seems to be a major cause. Ten of the world's seventeen penguin species already have been listed as threatened or endangered (Yoon 2001, F-1). Overfishing and oil spills also threaten penguins in some areas. Warming water depletes the penguins' food sources and causes many species to quit breeding, as large numbers of birds that would be reproducing die of starvation.

"If we get a series of intense El Niños, they're going to disappear," Patricia Majluf, a conservation biologist at the Wildlife Conservation Society, said of the colony of Humboldt penguins she studies, whose numbers also have been declining. "We lost half during one bad El Niño and these are very slow breeding birds" (Yoon 2001, F-1). El Niño conditions cause marked warming of many oceanic areas in which penguins live and breed. Warmer water also encourages the growth of toxic algae that constitute killing "red tides" in some of the same waters.

As the oceans have warmed during the last half of the twentieth century, winter sea ice no longer extends north of the South Shetland Islands. Pack ice now forms in the area only occasionally, about one or two of every six to eight years. The pack ice contains frozen diatoms, food for young krill. Without the usual ice pack, the penguins go hungry or starve. By 2001, ice had failed to form in many spawning areas six winters in a row, the life span of most krill. Given one more iceless winter, many of the aging krill will die, blowing a very large hole in the Adelie penguins' food chain.

David Pole Evans, a Falklands farmer, was the first to notice that something was amiss when he saw penguins "just standing around, not looking very fit or healthy" in April 2002. A few weeks later, he found thousands of dead penguins on the shore of Saunders Island and in the surrounding waters (Finch 2002, 12). Evans told the British Broadcasting Corporation that he estimated as many as 9,000 rockhopper penguins and 1,000 gentoo penguins, which are native to the South Atlantic region, had died within a short time.

Postmortem examinations failed to produce any firm reasons for the penguins' deaths. Nick Huin, the scientific officer from Falklands Conservation, a charity, said: "We're worried, because we don't know exactly why it's happening" (Finch 2002, 12). Tracking devices have been attached to ten penguins in an attempt to solve the mystery. But Mike Bingham, a researcher who works with the International Penguin Conservation Work Group, based on the islands, said there was "no doubt" that the penguins are dying from starvation (p. 12).

The local penguins have been molting a month later than usual. When they molt, they are no longer waterproof and therefore come ashore. Once removed from their food sources, the penguins need to have enough fat reserves to survive on land. One possible explanation for the deaths is overfishing. Another is a drop in the sea temperature caused by the melting of Antarctic glaciers as a result of global warming. The cooler waters have caused the penguins' food source of squid to remain in Argentine waters rather than move on the current to the Falklands (Finch 2002, 12).

Warming in the Southern Ocean during the last quarter of the twentieth century cut populations of emperor penguins living on the Antarctic coast in half "because of a decrease in adult survival during the late 1970s," an unusually warm period with reduced coverage of sea ice (Barbraud and Weimerskirch 2001, 185). A sudden temperature rise during the late 1970s and early 1980s coincided with a sharp drop in the survival rates of adult birds, according to Christopher Barbraud and Henri Wemerskirch writing in *Nature*.

Their research compared the size of a colony of penguins at Dumont d'Urville Station in Terre Adelie, Antarctica, with regional weather records from over fifty years. Average winter temperatures at the site rose from minus 17.3 degrees C in the early 1970s to minus 14.7 degrees C by the late 1970s, reducing the amount of local sea ice. According to a report in the *London Daily Telegraph*, during the same period, penguin numbers at the colony dropped to about 3,000. The death rate was higher for males than females. Although the population has since stabilized, there are concerns that other colonies may have suffered similar declines. "In years with high sea-surface temperatures emperor penguins probably have difficulties in finding food, which could increase mortality," the scientists reported (Derbyshire 2001, 6). Reduced coverage of sea ice also is associated with lower numbers of krill.

OTHER EXTINCTIONS AND WARMING IN ANTARCTICA

Penguins are only one of several Antarctic animals that could go extinct in coming years due to rapid habitat change provoked at least in part by warming temperatures. Global warming could wipe out thousands of Antarctic animal species in the next 100 years, the British Antarctic Survey asserted during 2002. An anticipated temperature rise of 2 degrees C, a fraction of what the Intergovernmental Panel on Climate Change (IPCC) forecasts by the end of the twenty-first century, would be enough to threaten large numbers of fragile invertebrates with extinction, said Professor Lloyd Peck from the British Antarctic Survey. These include exotic creatures found nowhere else on Earth, "such as sea spiders the size of dinner plates, isopods—relatives of the woodlouse, and fluorescent sea gooseberries as big as rugby balls" (Von Radowitz 2002). Peck said, "We are talking about thousands of species, not four or five. It's not a mite on the end of the nose of an elk somewhere" (2002).

Laboratory experiments have indicated that these animals and many others cannot survive even small variations in the temperatures of their habitats. Several thousand species of small animals, including mollusks and worms, are likely to die, including 750 species of sand flea alone. In time, fish populations and larger species such as penguins, seals, and whales further up the food chain could be affected, said Peck. He said it was impossible at this stage to guess how severe the consequences might be (Von Radowitz 2002). With a temperature rise of 2 to 3 degrees C, many Antarctic species' movements become sluggish as less oxygen reaches their tissues. With their ability to swim and feed compromised, their survival is threatened.

A CLIMATIC "MASTER SWITCH"

The Antarctic Peninsula has warmed several degrees C as its ice shelves disintegrate into the ocean, while parts of interior Antarctica are cooling as some of its glaciers thicken. An explanation of this apparent paradox may lie in a climatic "master switch" over the high southern latitudes, a circular wind pattern (the Antarctic Oscillation or Southern Annular Mode) that has been driven faster by the depletion of stratospheric ozone. According to one analysis, the cooling of inland Antarctica may reverse as the rupture in stratospheric ozone heals (Shindell and Schmidt 2004).

The work of David W. J. Thompson and Susan Solomon may be "the strongest evidence yet" that a shift in the Antarctic Oscillation "could explain a number of different components of [Antarctic] climate trends," according to David Karoly, a meteorologist at Monash University in Clayton, Australia (Kerr [May 3] 2002, 825). The researchers link cooling in the stratosphere induced by depletion of ozone levels to acceleration of the winds. "During the summer-fall season," Thompson and Solomon have written, "the trend toward stronger circumpolar flow has contributed substantially to the observed warming over the Antarctic Peninsula and Patagonia and to the cooling over eastern Antarctica and the Antarctic plateau" (Thompson and Solomon 2002, 895).

Writing in the May 3, 2002, edition of *Science*, David W. J. Thompson, a professor of atmospheric science at Colorado State University, and Susan Solomon, a senior scientist at the National Oceanic and Atmospheric Administration in Boulder, Colorado, asserted that ozone depletion over the Antarctic may help explain both contradictory trends. "Ozone seems to be capable of tickling the Southern Hemisphere patterns," Thompson said (Chang [May 3] 2002, A-16). Thompson and Solomon assert that a vortex of winds blowing around Antarctica that traps cold air at the South Pole has strengthened in the past few decades, keeping the cold air even more confined. The

Antarctic Peninsula lies outside the wind vortex and thus escapes the cooling effect. Ozone depletion "may" be a key causal factor in strengthening the wind pattern, according to Thompson and Solomon. "That's where we speculate," Dr. Thompson said, "and the emphasis is on the word 'may'" (p. A-16).

Scientists already knew that ozone depletion had cooled the upper atmosphere. Thompson and Solomon's research indicates that over parts of Antarctica the troposphere, the lowest six miles of the atmosphere, also has cooled. "It's a lot of food for thought in there," said Dr. John E. Walsh, a professor of atmospheric science at the University of Illinois (Chang [May 3] 2002, A-16). Walsh commented that the data tying the cooling to stronger winds was convincing. "My one reservation," he said, "is the link to the ozone" (p. A-16). He noted that the ozone hole was usually largest in November or December but that the greatest cooling was about six months later. Thompson agreed that ozone depletion could not explain the whole climatic picture and mentioned that other influences like ocean currents probably played important roles, too. "I seriously doubt it's the only player," he said. "I think it's one of many" (p. A-16).

The idea that stratospheric ozone depletion has been a factor in driving a stronger circumpolar vortex (and resulting cooling inside the vortex, with warming outside it) has been gaining support. In 2003, Nathan P. Gillett and David W. J. Thompson published results of a modeling study supporting this effect during the spring and summer. "The results," they wrote in *Science*, "provide evidence that anthropogenic emissions of ozone-depleting gases have had a distinct impact on climate not only at stratospheric levels but at Earth's surface as well" (Gillett and Thompson 2003, 273; Karoly 2003, 236–237).

Mark P. Baldwin and colleagues wrote in *Science*: "The resulting ozone 'hole' leads to a relative reduction in solar heating and a stronger vortex. Observations and recent model simulations show that the strengthening of the polar vortex during spring leads to lower surface temperatures over Antarctica and higher temperatures in the mid-latitudes of the Southern Hemisphere that persist into summer" (Baldwin et al. 2003, 317).

THE PAST AS PROLOGUE

Scientists have explored the paleoclimate of Antarctica in search of conditions similar to those forecast for the end of the twenty-first century as temperatures and atmospheric carbon dioxide levels rise. A study of Antarctica 24 million years ago, when carbon dioxide levels were very high compared to long-term averages, has revealed major expansions and contractions of Antarctica's ice, leading to worldwide sea-level changes of up to 200 feet.

T. R. Naish and colleagues have presented sediment data from shallow marine cores in the western Ross Sea that exhibit well-dated cyclic variations that link the extent of the East Antarctic ice sheet directly to orbital cycles during the Oligocene/Miocene transition (24.1 to 23.7 million years ago) (Naish et al. 2001, 719), a time when planetary temperatures averaged 3 to 4 degrees C warmer than today and the atmospheric carbon dioxide level was twice as high. These studies, they say, "should help to provide realistic analogues for their future behavior following the increased levels of atmospheric CO_2 and temperature projected for the end of this century" (p. 723).

An international research team studied sediment and fossil samples from 3,000 feet below the surface in the Cape Roberts area of the East Antarctica ice sheet, which covers most of the continent. The scientists focused on a 400,000-year period roughly 24 million years ago when carbon dioxide levels were twice as high as today and temperatures were up to 4 degrees C higher. These conditions are similar to those anticipated for the Earth in the next 50–100 years due to human-induced rises in greenhouse gas levels.

The team, whose study was published October 18, 2001, in *Nature*, stated: "Studies of Antarctic ice sheets during that time should help to provide realistic analogues for their future behavior following the increased levels of atmospheric CO_2 and temperature projected for the end of this century" (Dalton 2001, 7). Peter Webb, professor of geological sciences at Northern Illinois University, commented:

> To find out what we are in for, we had to go back to a time when there were similar conditions. We wanted to find a comparable

period in the past when carbon-dioxide levels were as high as they are expected to reach. There is every indication that this will be a model for what will happen in the future. We found the ice sheets 24 million years ago were smaller and more dynamic than today. There is good evidence for a lot of instability at the margins of the sheets. . . . We know there are glacial cycles, with the ice altering more when the climate is warmer, but it is now a question of working out how the speed and frequency of these cycles will change. They were probably faster in the past and will probably be faster in the future. (p. 7)

12 MOUNTAIN GLACIERS IN RETREAT

Mountain glaciers are in rapid retreat around the Earth, with very few exceptions. Climbers are being plucked from the Matterhorn in the Swiss Alps as thawing mountainsides crumble under them. During the summer of 2003, Mont Blanc, Europe's tallest, was closed to hikers and climbers because its deteriorating snow and ice was too unstable to allow safe passage. The mountain was crumbling as ice that once held it together melted during a record-warm summer in Europe. In the Swiss Alps, scientists have estimated that by 2025 glaciers will have lost 90 percent of the volume they contained a century earlier. Roger Payne, a director of the Swiss-based International Mountaineering and Climbing Federation, said global warming was emerging as one of the biggest threats to mountain areas. "The evidence of climate change was all around us, from huge scars gouged in the landscapes by sudden glacial floods to the lakes swollen by melting glaciers" (Williams 2002, 2).

Eighty-five percent of the glaciers in Spain's Pyrenees have melted during the twentieth century, according to Greenpeace, which reported: "The surface of the glaciers of the Pyrenees on the Spanish side went from 1,779 hectares (4,394 acres) in 1894 to 290 acres in 2000. . . . That infers a loss of 85 percent of the surface of the glaciers in the last century, with the process accelerating in the last 20 years" ("More Than" 2004). Glaciers in that area are expected to vanish by the year 2070. Melting glaciers are revealing a large number of previously buried historical artifacts. For example, a 450-year-old bison skull was found in a melting snowbank in the Colorado Rockies. Human cadavers, airplanes, dead birds, caribou carcasses, mining equipment, and prehistoric weapons have been uncovered (Erickson 2002, 6-A).

The snow and ice crown of Mount Kilimanjaro in equatorial Africa, made famous by Ernest Hemmingway a century ago, may vanish by the mid twenty-first century. Kilimanjaro will no longer live up to its name, which in Swahili means "mountain that glitters." Mount Kenya's ice fields have lost three-quarters of their entire extent during the twentieth century (Lynas [March 31] 2004, 12). By the end of the twenty-first century, Glacier National Park in Montana may lose the last of its permanent glaciers; its name will be a daily reminder of what humankind has done to the Earth's climate. The original 150 glaciers within Glacier National Park had been reduced to thirty-seven by 2002, and most of these were small remnants of once-mighty ice masses. Glaciers in New Zealand's Southern Alps lost 25 percent of their surface area during the last century (McFarling [September 25] 2002, A-4).

THE WORLDWIDE STATE OF GLACIERS

The U.S. Geological Survey has collected a digital library that describes the state of more than 67,000 glaciers around the world. Using historical photographs, images from space satellites, precision laser measurements, and other tools, the archive tells a story of glacial retreat around the world (Toner [June 30] 2002, 4-A). "The rate of ice loss since 1988 has more than doubled," said Mark Meier, a geologist at the University of Colorado at Boulder. "Some glaciers around the world are now smaller than they have been in the last several thousand years" (p. 4-A). Glaciologist Mark Dyurgerov of the University of Colorado's Institute of Arctic and Alpine Research estimated that mountain glaciers around the world are losing about twenty-five cubic miles of ice each year due to climate change (Erickson 2002, 6-A). Worldwide, the "equilibrium-line altitude" of the average glacier—the point below which it loses mass on an annual basis due to melting—rose in elevation an average of about 200 meters between 1960 and 1998 (Lynas [*High Tide*] 2004, 233).

Generally, the only glaciers gaining mass are in wet maritime areas of the world, such as parts of Norway and Sweden, where melting has been offset by increased snowfall, another facet of climate change. Alaska's Hubbard Glacier "is advancing so swiftly that it threatens to seal off the entrance to Russell Fiord near Yukatat and turn the fiord into an ice-locked lake. Like a handful of other Alaska glaciers, the Hubbard is fed by a high-altitude snowfield that has not yet been affected by warmer temperatures" (Toner [June 30] 2002, 4-A). "The Hubbard is definitely an exception," said the Geological Survey's Bruce Molnia, who has been tracking 1,500 Alaska glaciers. "Every mountain group and island we have investigated is seeing significant glacier retreat, thinning or stagnation, especially at lower elevations. Ninety-nine percent of the named glaciers in Alaska are retreating" (p. 4-A).

Scientists reported during October 2003 that the Patagonian ice fields of Chile and Argentina have been thinning so swiftly that this 6,500-square-mile region of South America now exhibits a pace of glacial retreat that is among the most rapid on Earth. During the period 1995–2000, rate of volume loss for sixty-three glaciers in the area doubled, compared to the 1968–2000 average (Rignot, Rivera, and Casassa 2003, 434). Early in 2004, Greenpeace International released results of an aerial survey confirming the rapid recession of the Patagonia glaciers, which was estimated at forty-two cubic kilometers a year, an amount that could fill a large sports stadium 10,000 times. Greenpeace compared photographs taken during January 2004 with images from 1928, showing that "the glaciers had significantly thinned and retreated several kilometers" (Hodge 2004, 8).

"These losses are not just regrettable but actually threaten the health and well-being of us all. Mountains are the water towers of the world, the sources of many rivers. We must act to conserve them for the benefit of mountain people, for the benefit of humankind," said Klaus Toepfer, head of the U.N. Environment Programme (Vidal [October 24] 2002, 7). Glaciologist Lonnie G. Mosley-Thompson, a senior research scientist at the Byrd Polar Research Center of Ohio State University, said that the rate of glacial retreat is most pronounced in equatorial areas, such as along the spine of the Andes, where hydrological consequences may be the most significant.

Late in 2003, the World Wildlife Fund (WWF) said that millions of people around the world will face severe water shortages as glaciers around the world melt unless governments take urgent action to deal with global warming. "Increasing global temperatures in the coming century will cause continued widespread melting of glaciers, which contain 70 percent of the world's fresh water reserves. . . . An overall rise of temperature of four degrees Celsius before the end of the century would eliminate almost all of them," WWF warned ("Billions" 2003). Ecuador, Peru, and Bolivia, where major cities rely on glaciers as their main source of water during dry seasons, would be worst affected, the WWF forecast. Areas of the Himalayas face grave danger of flooding, the group said, noting that glacier-fed rivers in the region supply water to one-third of the world's population, mainly in India and China.

Chinese glaciologist Yan Tandong estimated that during the past four decades China's glaciers have shrunk an average of 7 percent per year, an amount equivalent to all the water in the Yellow River. Yao previously told local media that as many as 64 percent of China's glaciers (mostly in Tibet) may be gone by 2050 if current trends continue. The human cost could be immense, since 300 million Chinese live in the country's arid west and depend on water from the glaciers for their survival ("Global Warming Makes" 2004).

Temperatures are rising quickly at higher elevations. For example, David L. Naftz and colleagues from the U.S. Geological Survey reconstructed trends in air temperature from alpine areas of the Wind River Range in Wyoming and found, from ice-core samples, that temperatures had increased about 3.5 degrees C from 1960 to the early 1990s. Furthermore, temperatures had increased about 5 degrees C from the end of the "Little Ice Age" about 1850 to the early 1990s. Naftz and colleagues wrote in the *Journal of Geophysical Research* that their readings were "in agreement with [temperature] increases observed at selected high-altitude and high-latitude sites in other parts of the world" (Naftz et al. 2002).

DISINTEGRATION OF GLACIERS IN THE HIGH ALPS, AND ELSEWHERE IN EUROPE

Rising temperatures have been melting ancient glaciers on the high Alps. This melting has brought devastating rockslides in the summer that have endangered the lives of many climbers, including seventy in one day, July 14, 2003, which resulted in one of the largest mass rescues in the area's history. Most were plucked from the Matterhorn, which was racked by two major landslides that day. According to one observer, "Those climbing its slopes could have been forgiven for thinking the crown jewel of the Alps had started falling apart under their feet" (McKie 2003, 18).

According to Robin McKie, writing in London's *Observer*, "The great mountain range's icy crust of permafrost, which holds its stone pillars and rock faces together, and into which its cable car stations and pylons are rooted, is disappearing" (2003, 18). Several recent Alpine disasters, including the avalanches that killed more than fifty people at the Austrian resort of Galtur during 1999, have been blamed on melting permafrost. During August 2003, the freezing line in the Alps rose to 13,860 feet (4,200 meters), almost 4,000 feet above its usual summer maximum of 9,900 feet (3,000 meters) (Capella 2003).

Scientists attending the 2003 International Permafrost Association conference in Zurich, Switzerland, said that conditions in the high Alps could get more dangerous in coming years, assuming continued warming. "I am quite sure what happened on the Matterhorn last week was the result of the Alps losing its permafrost. We have found that the ground temperature in the Alps around the Matterhorn has risen considerably over the past decade. The ice that holds mountain slopes and rock faces together is simply disappearing. At this rate, it will vanish completely—with profound consequences," said civil engineer Professor Michael Davies of Dundee University, a conference organizer (McKie 2003, 18).

Air temperature increases in the Alps are being amplified fivefold underground. A borehole dug at Murtel in southern Switzerland has revealed that frozen subsurface soils have warmed by more than 1 degree

C since 1990. In addition to general air temperature rises that are heating the ground, increased evaporation caused by warmer summers has triggered heavier snows, which insulate the soil and keep it warmer in winter. Ice also becomes more unstable as it warms, raising the danger of devastating landslides.

Melting of permafrost in the Alps and other European mountain ranges does much more than spoil mountain climbers' treks. Alpine villages and ski resorts are sometimes threatened with devastating landslides. Fear has been expressed that some villages may have to be evacuated. Rivers might be blocked by debris, causing flash floods when these unstable dams subsequently collapse. According to a report in the *London Guardian*, Charles Harris of the Earth Sciences Department at Cardiff University, who coordinates research for the European Union, said that the main areas at risk are the Alps in Switzerland, Austria, France, Germany, and Italy, where the mountains are densely populated and the slopes are very steep. According to this account, among the places being monitored is the Murtel-Corvatsch Mountain above fashionable St. Moritz, and the Schilthorn, which towers above the Muran and Gandeg resorts near Zermatt (P. Brown [January 4] 2001, 3). Harris said that the Swiss Alps had warmed by 0.5–1.0 degrees C during the past fifteen years (Clover 2001, 12).

A new organization called Permafrost and Climate in Europe (PACE) has been established to monitor the effects of climate change on the stability of mountains. PACE literature contends that "the combination of ground temperatures only slightly below the freezing point, [along with] high ice contents and steep slopes, makes mountain permafrost particularly vulnerable to even small climate changes" (Brown [January 4] 2001, 3).

FLOOD THREATS BELOW THE ALPS

Melting glacial ice has been increasing the volume of the Rhine, Rhone, and Po rivers. Since 1850, the volume of Europe's glaciers has shrunk by 50 percent (Toner [June 30] 2002, 4-A). The great rivers of Europe that flow from the Alps may decline, however, from swollen summer torrents to mere trickles as ancient mountain ice fields disappear during the twenty-first century due to global warming. David Collins of Salford University presented findings to a Royal Geographical Society conference in Belfast indicating that the ice that is now melting in the Alps accumulated during the "Little Ice Age" between the fifteenth and eighteenth centuries. "The [present] combination of warmer summers and drier winters, meaning less snow to feed the glaciers, has meant that the vast bank of ice on the mountain tops is disappearing," said Collins. "The ice is like money in the bank, if you keep drawing more than you put in, eventually it runs out" (P. Brown [January 5] 2002, 9).

Before the rivers dry up, warming will supply a final torrent as the glaciers melt. The summer flows of the rivers fed from the Alps, including the Rhine, Rhone, Po, and Inn (which feeds the Danube), have recently been higher than they have been for centuries. The excessive water flow has been good news for those living on riverbanks in southern and eastern Europe who draw off the excess water for irrigation and domestic use. In France, the river water was used for cooling nuclear power stations.

Forecasts by Collins and his team showed that these boom times for water supply will soon end. When all the ice goes, the summer flow of the rivers will be almost entirely dependent on rainfall. Under some climate models, rainfall in southern Europe probably will decline even further under warmer conditions. "We can see serious potential problems but it is very hard to be precise because weather patterns could change again. . . . Some of the glaciers—for example, there are a number of small ones at Gornergrat, near Zermatt—are now below the snow line in summer. This means they are doomed. The ice they are made of was laid down in snowfall two or three centuries ago and is melting away faster each year," Collins reported (P. Brown [January 5] 2002, 9). Collins reported that the reduction in the glaciers of the Alps had been matched by an increase in glacier size in the Jotunheimen Range in Norway, because increased precipitation in northern Europe, in this case falling as snow, had blanketed those glaciers and protected them from any temperature increases. This increased precipitation had led to a net increase in the size of glaciers over the same period in that area, just as those in the Alps were retreating.

The people of Macugnaga (pronounced maa-COON-yaga), Italy, an Alpine resort village, long ago learned to cope with the floods that sometimes accompany the melting snow in the spring, "but nothing," according to one account, "prepared them for the catastrophic flood threat they now face—a glacier rapidly melting from unusually warm temperatures" (Konviser 2002, C-1). During July 2002, as many as 300 officials and volunteers struggled under a state of emergency "to prevent a gigantic glacier-fed lake from breaking through the giant ice wall that confines it" (p. C-1). If they failed, a devastating wall of water carrying chunks of glacier and mountainside could surge through this verdant valley. Known technically as a "glacier lake outburst flood," or GLOF, "it's an event previously seen only in the Himalayas where the slopes of the mountains are steeper. But scientists say the threat is both real, and a warning of things to come if the global-warming trend continues" (p. C-1).

"It's a dangerous situation because the border of the lake is ice, which isn't stable," said Claudia Smiraglia, a professor of physical geography at Milan University. "The glacier is always in motion" (Konviser 2002, C-1). Bruce I. Konviser, reporting for the *Boston Globe*, described the potential scope of the threat: "If the water escapes, the 650 residents of Macugnaga and as many as 7,000 vacationers, depending on the time of year—would have approximately 40 minutes to gather their belongings and get to higher ground before the wave of water and mountain wipes out much, if not all, of the manmade structures, according to Luka Spoletini, a spokesman for the Italian government's Department of Civil Protection" (p. C-1).

GLACIER NATIONAL PARK WITHOUT GLACIERS?

When the 1.4 million-acre Glacier National Park was created early in the twentieth century, it included more than 150 glaciers in rugged crags and valleys along northern Montana's continental divide. A hundred years later, only thirty-seven of them remained. By the middle

People view an ice cave in Boulder Glacier at Glacier National Park, Montana, July 1932 (top). The same area is shown in a 1988 photo at bottom. © AP/Wide World Photos.

of the twenty-first century, given present trends, the park with "glacier" in its name will have no permanent ice (Toner [June 30] 2002, 4-A).

The Grinnell Glacier, for example, has been retreating more than fifteen feet a year. The remaining glaciers in 2002 cover less than a third of the area they occupied when the park was created in 1910. The pace of melting has been accelerating with the advent of the new century. "Even those of us who work on these glaciers are surprised at how quickly they are melting," said Dan Fagre of the U.S. Geological Survey. "At the current rate, we may see the complete disappearance of functioning glaciers in the park within 30 years" (Toner [June 30] 2002, 4-A). "Glaciers don't know anything about global warming or its causes, but they are excellent barometers of climate change," Fagre said. "They integrate temperature, solar radiation and snowfall and express it as a big lump of ice. They don't have a political agenda. They just reflect what's happening" (p. 4-A).

Spring has been arriving earlier in the park; summers are warmer and longer. Most years, more ice melts than the winter snows replace. During the winter, precipitation now falls more often as rain rather than snow, even at relatively high altitudes (Toner [June 30] 2002, 4-A).

ROCKY MOUNTAIN GLACIERS: GONE IN THIRTY YEARS?

The glaciers of the Rockies, a major source of water for the western half of North America, will be gone in twenty to thirty years, according to David Schindler, a University of Alberta ecologist who holds Canada's top science prize—the $1 million Gerhard Herzberg Gold Medal. Schindler, also a renowned expert on water quality, asserted that the situation is so dire it cannot be reversed. "This is the bad news. Because we've procrastinated for the past 25 years, we're locked into considerable warming for the next several decades, and I think that will be enough to finish off the glaciers of the Rockies," he said. With a concerted effort, the glaciers could recover in the long term. "Right now we are getting more melting than precipitation," Schindler said.

"That could be reversed with a cooler climate. I think that, looking ahead a century, we could get them back. All we need to do is reverse the balance" (Remington 2002, A-19).

"When those glaciers go, I wonder what we will be doing for municipalities and agriculture on the southern [Canadian] prairies. Those rivers are going to be streams at best once climate warming takes its toll on water supply," said Schindler (Remington 2002, A-19). "The sorts of melting phenomena we see are everywhere—the Alps, the Rockies, the mountains of Africa."

Glacial melt water is released slowly, allowing streams and rivers to flow throughout the year, including mid- to late summer when water demand reaches its annual peak. Should glaciers disappear, people will have to rely on trapping water when the snow melts in the spring. "I know what people will say: 'Build more water reservoirs and trap more spring flow.' That doesn't bode very well for people downstream," Schindler said (Remington 2002, A-19).

Schindler has been analyzing climate records for Canadian prairie cities. He has found temperature increases ranging between 1 and 4 degrees C since recordkeeping began in the early twentieth century. "These can't be dismissed as urban heat-island effects, as a lot of the debunkers would like to say. The biggest increase of all is Fort Chipewyan [an isolated community in northeastern Alberta]. There is certainly no heat island effect in Fort Chipewyan," Schindler said (Remington 2002, A-19).

Ice patches along the continental divide in Colorado have disgorged ancient bison horns that have been radiocarbon dated between 2,090 and 2,280 years old, suggesting that, in some cases, ice along the divide has retreated to levels unseen since before the time of Christ (Erickson 2004). "Over the last couple of decades, and especially over the last 10 years, we have entered a period of warming and retreat that is as great, or greater, than any we know of since the end of the last ice age [10,000 years ago]," said glaciologist Tad Pfeffer, of Colorado University's Institute of Arctic and Alpine Research. "The Front Range glaciers and snowfields could be gone in a couple of decades" (2004).

DROUGHT ANTICIPATED IN ALBERTA

In Canada's province of Alberta, experts have warned that the area is headed for a massive drought worse than the "dust bowl" conditions of the 1930s, in large part because of dwindling supplies from mountain snowpack coupled with a rising number of people and livestock using water. "I see a disaster shaping up in Alberta and it's not a question of if, it's a question of when," said University of Alberta ecologist David Schindler, a water researcher and longtime critic of Alberta's water policies. "There is going to be a major drought coinciding with global warming and record numbers of people and livestock on the landscape. It's something that gives me nightmares" (Semmens 2003, A-14). Glacial cover in the Canadian Rockies is nearing its lowest point in at least 10,000 years. Statistics Canada reported that 1,300 glaciers in the country have lost between 25 and 75 percent of their mass since 1850. Most losses have been recorded in the last fifty years. The report said that most of these losses can be attributed to global warming (Paraskevas 2003, A-8).

Schindler cited research by several Alberta scientists indicating that the average temperature for western Canada is the highest it has been in at least 10,000 years and is expected to continue climbing. Such conditions will lead to continued recession of glaciers in the Rocky Mountains, with less rain and snowfall as well as increasing evaporation of water. The flow of some of Alberta's rivers is already running up to 80 percent lower than 100 years ago, and Schindler said Alberta is overdue for a ten-year drought that tends to hit the province every 100 to 150 years (Semmens 2003, A-14).

THE DANGERS OF GLACIAL RETREAT
IN THE HIMALAYAS

In the Himalayas, the Rongbuk Glacier on the north face of Mount Everest retreated between 170 and 270 meters from 1966–1997. The glacier from which Sir Edmund Hillary and Tenzing Norgay set out to climb Mount Everest nearly a half-century ago has retreated about three miles up its host mountain. Much of that glacier has turned to melt water, according to U.N. observers who visited the site (Williams 2002, 2). Roger Payne, one of the observation team's leaders, said, "Back in 1953 when Hillary and Tenzing set off to climb Everest they stepped out of their base camp and straight on to the ice. You would now have to walk for over two hours [from the same site] to get on to the ice" (p. 2). "Glaciers in the Himalayas are wasting at alarming and accelerating rates, as indicated by comparisons of satellite and historic data, and as shown by the widespread, rapid growth of lakes on the glacier surfaces," said Jeff Kargel of the U.S. Geological Survey, international coordinator for Global Land Ice Measurements from Space (GLIMS) ("Glacial Retreat" 2002).

Melt water from the Himalayas' Imja Glacier, ten kilometers (six miles) to the east-southeast of Mount Everest, has created a vast lake held back only by an unstable natural dam comprised of boulder debris that once marked the edge of the glacier. A collapse of this dam could send a wall of water up to 100 meters high surging down the valley, which is inhabited by the Sherpas, who have assisted climbing expeditions' ascents of Mount Everest. The valley is also the main approach route to the Everest Base Camp. "We know it's going to go shooting down the flood plain, and in a mountainous area, that's where the people live," said Lisa Graumlich, who directs the Big Sky Institute at Montana State University (McFarling [September 25] 2002, A-4).

The Dig Tsho glacial outburst in Nepal during 1985 destroyed a hydroelectric plant, wiped out fourteen bridges, and drowned dozens of villagers. The danger is so obvious, Graumlich said, that some Himalayan villages have installed primitive warning systems—basically a system of horns—in an attempt to save lives during the next flood (McFarling [September 25] 2002, A-4). "We're just watching [glacial lakes] form in the Himalayas and Peru," said Alton C. Byers, director of research and education for the West Virginia—based Mountain Institute. "All you have to do is release that dam and you'll lose vast amounts of water in seconds" (p. A-4).

By 2002, the snout of the Himalayan glacier that feeds the mighty Ganga (Ganges River) had developed giant fractures and crevices along a ten-kilometer stretch, indicating massive ice melting. During fifteen years of researching such phenomena, Syed Iqbal Hasnain, who heads the Glacier Research Group at Delhi's Jawaharlal Nehru University, had never seen such a rapid deterioration of the frozen massif. He commented, "If the rate continues, we could see much of the Gangotri glacier and others in the Himalayas vanish in the next couple of decades" (Chengappa 2002, 40). Hundreds of millions of people who live within the watershed of the Ganges depend on the water shed by Himalayan glaciers to some degree. Half of India's hydroelectric power is generated from glacial runoff. Ninety percent of the water used in desert areas of Pakistan is from the River Indus, supplied by glaciers that have been rapidly losing mass for most of the twentieth century (Lynas [*High Tide*] 2004, 238).

More than forty lakes high in the Himalayas that have formed from rapidly melting glaciers are expected to burst their banks before the year 2010, sending millions of gallons of water and rocks cascading onto towns and villages. These lakes are growing larger and more unstable as temperatures rise. As many as forty-four such lakes have been identified, twenty in Nepal and twenty-four in Bhutan. There are thought to be hundreds more such "liquid time bombs" in India, Pakistan, Afghanistan, Tibet, and China. Surendra Shrestha, Asian regional coordinator for the U.N. Environment Programme's early-warning division, said: "These 44 could burst their banks with potentially catastrophic results for people and property hundreds of kilometers downstream" (P. Brown [April 17] 2002, 13). Nepal's Tsho Rolpa Lake has grown sixfold in size since the 1950s and is now 2.6 kilometers long, 500 meters wide, and as deep as 107 meters. A flood from this lake could cause serious damage as far as 108 kilometers downstream in the village of Tribeni, threatening 10,000 lives (p. 13).

Temperatures in the region have risen by 1 degree C since the 1950s, causing thousands of glaciers to retreat by an average of thirty meters a year. The worst recorded collapse of one of these dams occurred in 1954, when 300,000 cubic meters of water and rock poured without warning into China in a forty-meter-high flood surge from the Sangwang dam on the Tibet-Nepal border. The city of Gyangze, 120 kilometers away, was destroyed. The dead totaled many thousands.

"Glaciers have never been retreating at this rate, and lakes have never been forming so quickly," Shrestha said. "For most of the Himalayas the risk is unknown, but it is very great. This is a seismic zone, and a catastrophic flood could be sparked by an earthquake" (P. Brown [April 17] 2002, 13).

THE QINGHAI-TIBET RAILWAY AND DETERIORATING PERMAFROST

Designers and builders of the Qinghai-Tibet Railway have found that deteriorating permafrost has presented problems for the world's highest-elevation rail line. Research produced by the Chinese Academy of Sciences indicates that permafrost on the path of the railroad is now five to seven meters thinner than twenty years ago; about 10 percent of the plateau's permafrost has completely melted. Scientists said the climate changes have changed ground temperatures to a depth of at least forty meters, with greatest changes to a depth of twenty meters. Stabilizing the melting earth has become a major challenge in designing the railway. Engineers working on the railway have adopted three special measures to ensure the stability of the roadbed in the permafrost areas, including changing routes, building railway bridges along sections of complex geological conditions, and building soil layers that can insulate the ground from heat created by the railway ("Global Warming Troubles" 2003).

The nearly 2,000-kilometer railway links Xining, the capital of northwest China's Qinghai Province, and Lhasa, the capital of the Tibet Autonomous Region. The section linking Xining and Golmud City in Qinghai was completed in 1984. Construction of the 1,118-kilometer section connecting Golmud with Lhasa began in June 2001 and is expected to be completed by 2007 ("Global Warming Troubles" 2003).

THE MELTING SNOWS OF KILIMANJARO

By 2002 Mount Kilimanjaro had lost 82 percent of its icecap's volume since it was first carefully measured in 1912, a third of it since 1990,

Aerial views of Mount Kilimanjaro, showing reduction in ice cap, 1993 and 2000. Courtesy of the Goddard Institute of Space Studies/NASA.

according to glaciologist Lonnie Thompson. Kilimanjaro's ice field shrank from 12 square kilometers in 1912 to only 2.6 square kilometers in 2000, reducing the height of the mountain by several meters. The ice covering the 19,330-foot peak "will be gone by about 2020," said Thompson (Arthur 2002, 7). The reduction of glacial mass already has cut water volume in some Tanzanian rivers that supply villages near the mountain's base.

Global warming may not be the only culprit in the demise of Kilimanjaro's icecap; natural climate changes also have been blamed, along with deforestation on the mountain's slopes, which sucks out of the rising winds the moisture that once coated the upper elevations of the mountain with snow. Euan Nisbet of Zimbabwe's Royal Holloway College has suggested, in all seriousness, that plastic tarps be draped across the remaining ice fields to extend their lives (Morton 2003).

The demise of Kilimanjaro's icecap could imperil Tanzania's economy, which relies on tourism driven by the attraction of the mountain. In the Hemingway short story "The Snows of Kilimanjaro," a disillusioned writer, Harry Street, reflected on his life while lying injured in an African campsite. The short story was made into a film starring Gregory Peck in 1952. "Kilimanjaro is the number-one foreign currency earner for the government of Tanzania," said Thompson. "It has its own international airport and some 20,000 tourists every year" (Arthur 2002, 7).

The Kilimanjaro icecap, which once was fifty meters deep, originated during an extremely wet period about 11,700 years ago, according to ice cores examined by Thompson and colleagues. Their research indicates that Kilimanjaro has lost its icecap before, during catastrophic droughts 8,300, 5,200, and 4,000 years ago. The most recent drought, which lasted 300 years, also threatened the rule of the Egyptian pharaohs. Professor Thompson explained, "Writings on tombs 'talk' about sand dunes moving across the Nile and people migrating. Some have called this the Earth's first dark age.... Whatever happened to cause these dramatic climate changes could certainly occur again" (Arthur 2002, 7).

Some scientists assert that the melting snows of Kilimanjaro are a result mainly of declining moisture rather than rising temperatures. A study published in the *International Journal of Climatology* found no evidence of temperature rise in the region. Researchers led by Georg Kaser, professor of tropical glaciology at the University of Innsbruck, Austria, and Douglas Hardy of University of Massachusetts have pinpointed a sharp drop in atmospheric moisture in the 1880s and a subsequent prolonged dry period in the region as behind the icecap's decline (Kaser et al. 2004, 329). When Kaser's findings were picked up by climate skeptics as "proof" that global warming has not contributed to the demise of Kilimanjaro's snows, however, he repudiated their assumption. "We are entirely against the black-and-white picture that says it is either global warming or not global warming," Kaser told the *New York Times*. "We have a mere 2.5 years of actual field measurements from Kilimanjaro glaciers ... so our understanding of their relationship with climate ... is just beginning to develop," said Douglas R. Hardy, team co-leader of the paper (Revkin [March 23] 2004).

Rapid melting of glaciers often blesses farmers downstream with a large amount of water in the short term. Farmers near Mt. Kilimanjaro, for example, recently found the water supply so bountiful that they have been able to grow a surplus of crops for sale. They have begun to raise water-hungry crops such as tulips for export to Europe. The water supply is a temporary blessing, of course. After the glaciers have melted, the present deluge of water will be replaced by severe drought. "More water now means more agriculture," Graumlich said. "But what will they do when there is much less water later on?" (McFarling [September 25] 2002, A-4).

DANGERS OF GLACIAL COLLAPSE IN RUSSIA

On September 20, 2002, a large avalanche in southern Russia killed as many as 150 people in North Ossetia, a small republic in the mountains near the Georgian border. According to an account from the scene, "A chunk of a glacier about 500 feet high broke off from beneath a mountain peak and roared down two gorges at more than 62 miles per hour, uprooting trees and accumulating mud and rocks as it went," burying much of the village of Karmadon ("100 Still Missing" 2002,

A-5). "Within minutes," according to another account, "an immense glacier severed from the mountain near its 15,700-foot peak, roared along the nearby hillside village of Karmadon, burying much of it up in as much as 500 feet of ice and debris.... The collapse of the Maili glacier on the northern edge of the Caucasus Mountains ripped out trees and tossed massive trucks as if they were toys. It left a 20-mile path of rocky debris, blackened ice and devastation" (Wines 2002, A-11). The avalanche eroded one-third of the Maili Glacier, about 3 million tons of ice. A team of experts was sent to the region to try to discover the cause of the disaster, which an Interior Ministry spokesman said might be connected to global warming. Among the dead was Sergei Bodrov Jr., one of Russia's best-known young movie stars, who was filming in the area when the avalanche hit.

Russian officials said that the collapse of the glacier seemed at least partly related to climatic change. The issue is tricky, because the collapse of glaciers can depend on a variety of factors, including temperature, rainfall, humidity, angle of slope, and even the reflectivity of the glacial ice. U.S. experts commented that the Maili Glacier incident followed the pattern of glacial collapse in other areas affected by rising temperatures. "Glaciers tend to [collapse] like that when they're receding, and glaciers are receding all over the world," said Dan Fagre, an ecologist and expert on the ramifications of glacier loss at Glacier National Park in Montana. Glacial collapses can be prompted by the accumulation of melt water, which often pools within the cracks of receding glaciers behind walls of sediment and stone, building pressure that eventually produces a slide of ice, mud, and rock (McFarling [September 25] 2002, A-4).

COMPARING PHOTOGRAPHS TO GAUGE GLACIAL RETREAT

Amanda Phelan of the *Sydney Sunday Telegraph* described a unique European project that uses old photographs, compared with present-day vistas, to document how climate change is affecting glaciers. "Two photographs, taken only 84 years apart, show how the Blomstrandbreen glacier on a remote island off Norway was reduced from a towering force, dominating the skyline, to little more than a wall of ice rubble. Less than a century on, photographs show the glacier on Svalbard, an island 603 kilometers off Norway, is a pitiful slope—proof that global warming is a threat to our planet, says Greenpeace, which took the second image this year" (2002, 47).

German geologists, in association with Greenpeace, used old picture postcards to measure the retreat of glaciers in Europe between 1850 and 2000. A Munich-based team of geological climatologists rummaged through antique shops, markets, state archives, and university libraries for old photos and postcards, collecting 2,500 items. They then hiked to the edges of the glaciers on the photographs and measured changes. They found that in a century and a half, "the giant ice fields shrank about a third in area and lost about half of their volume. In the last 25 years, they have melted even more quickly, losing an additional 20 to 30 per cent of their water content" (Hall 2002, 8).

The geologists spent months hiking through the Alps, sometimes with local climbers. They studied elevation maps, examined satellite images, and measured paths and meadows. Equipped with the most modern surveying technology such as gauges measuring ultraviolet light on the one hand and historic hiking maps on the other, the scientists were able to photograph the sixty largest Alpine glaciers pictured on the postcards. "We had to do a lot of climbing and clambering," said the team's project leader, Wolfgang Zaengl. "The postcards were our most invaluable tool. . . . We are witnesses to the fastest glacial melting in a thousand years. But today we are lucky that we can still see the glaciers. Future generations probably will not" (Hall 2002, 8). Results of their research have been published on the Internet at www.gletscherarchiv.de.

ANDES GLACIAL RETREAT

Hundreds of Andean glaciers are retreating, and scientists say that their erosion is a direct result of rising temperatures. During three decades (1970–2000), Peru's glaciers lost almost a quarter of their 1,225-square-mile surface (Wilson 2001, A-1). The 18,700-foot-high Quelccaya icecap in the Andes of southeastern Peru has been steadily shrinking at an accelerating rate, losing 10 to 12 feet a year between 1978 and 1990, up to 90 feet a year between 1990 and 1995, and 150 feet a year between 1995 and 1998. The glacier retreated between 100 and 500 feet, depending on location, between 1999 and 2004. The Peruvian National Commission on Climate Change forecast in 2005 that Peru will lose all its glaciers below 18,000 feet in ten years. Within forty years, the commission said that all of Peru's glaciers will be gone (Rergaldo 2005, A-1).

The Quelccaya icecap in the Peruvian Andes shrank from twenty-two to seventeen square miles between 1974 and 1998 and is retreating more than 500 feet a year (Nash 2002, 303). Water from hundreds of glaciers in a stretch of the Andes known as the Cordillera Blanca ("White Range") drives the rural economy of Peru. The water runoff moistens wheat and potatoes along the mountain slopes. It also lights the houses and huts with electricity generated by a hydroelectric plant on the river (Wilson 2001, A-1). Lima, a city of 8 million people situated in the Atacama, one of the driest deserts on Earth, receives nearly all of its water from glacial icemelt during a six-month dry season. Within a few decades, at present melting rates, Lima's people will encounter severe water shortages. The same water is used to generate much of Lima's electricity. At the same time as its few wells are drying up and glaciers are shrinking, Lima has been adding 200,000 residents a year. Bolivia's capital, La Paz, and Ecuador's Quito face similar problems (Lynas [*High Tide*] 2004, 236–237). Within a few years, the life-giving waters may diminish to a trickle if current trends continue.

Icemelt in Peru also may make some areas more vulnerable to the frequent earthquakes that afflict the area. "Glaciers usually melt into the rock, filling in fissures with water that expands and freezes when the temperatures drop. What scientists fear is that, with increased melting, more water and larger ice masses are pulling apart the rock and making the ice cap above more susceptible to the frequent seismic tremors that rock the area" (Wilson 2001, A-1).

Many Peruvians who face drought in the long term are being lulled into a sense of plenty in the short term by increasing glacial runoff.

The Jacamba Glacier in the Peruvian Andes, 1980 and 2000. Courtesy of Mark Lynas/Photos by Tim Helwig-Larsen.

According to Scott Wilson, writing in the *Washington Post*, the short-term glacial runoff "has made possible plans to electrify remote mountain villages, turn deserts into orchards and deliver potable water to poor communities. In some mud-brick villages scattered across the valley, new schools will open and factories will crank up as the glacier-fed river increases electricity production" (Wilson 2001, A-1).

"In the long run . . . these long-frozen sources of water will run dry," said Cesar Portocarrero, a Peruvian engineer who worked for Electroperu (the government-owned power company) and who has monitored Peru's water supply for twenty-five years. In the meantime, evidence of changing climate has appeared in Portocarrero's hometown, Huaraz, a small city at 10,000 feet in the Andes. "I was doing work in my house the other day [in 2002] and saw mosquitoes," Portocarrero said. "Mosquitoes at more than 3,000 meters. I never saw that before. It means really we have here the evidence and consequences of global warming" (Revkin [August 28] 2002, A-10).

Benjamin Morales, the dean of Peru's glaciologists, calls the glaciers of Peru (80 percent of the world's tropical icepack) "the world's most sensitive thermometers," because they react to the smallest changes in temperatures. Many of the Peruvian glaciers that are now melting formed more than a million years ago. Today, some of these glaciers are melting so quickly that "residents have watched a usually painstakingly slow geological process with their own eyes. Since 1967, when Peru began monitoring its glaciers, scientists estimate, the ice caps have lost 22 percent of their volume, enough to fill more than 5.6 million Olympic-size swimming pools" (Wilson 2001, A-1).

Glaciologist Lonnie Thompson has estimated that many of Peru's glaciers could disappear during the next fifteen years (Wilson 2001, A-1). In Bolivia, the mass of glaciers and mountain snowcaps has shrunk 60 percent since 1978, raising a specter of water shortage for La Paz, home of 1.5 million people, as well as nearby El Alto (Forero 2002, A-3).

For twenty-five years, Thompson has been tracking a particular Peruvian glacier, Qori Kalis, where the pace of shrinkage accelerated enormously during the late 1990s. From 1998 to 2000, the glacier receded an average of 508 feet a year, according to Thompson. "That's thirty-three times faster than the rate in the first measurement period," he said, referring to a previous study of the glacier that covered the years 1963–1978 (Revkin 2001, A-1).

NEW ZEALAND GLACIERS SHRINKING

In New Zealand, as in many other mountainous areas, South Island rivers are running with full banks from melting glacial waters. National Institute for Water and Atmospheric Research consultant glaciologist Trevor Chinn said that many of the South Island's largest glaciers have been retreating rapidly. As they have melted, the glaciers have been adding millions of gallons of water to local rivers. "Current flows are higher than what you would expect from rainfall. What we are seeing is borrowing water from glacier storage for the Waitaki and Clutha rivers," Chinn said (Robson 2003, 13). A 0.5–3.0 degree C rise in average global temperature would reduce New Zealand's glacial mass by 25–50 percent, according to one estimate. The glaciers have lost 20 percent of their area in the past 100 years, and many larger ones are collapsing into lakes.

2002: GREAT LAKES AND NEW YORK LAKES FAIL TO FREEZE

The Great Lakes failed to freeze during the winter of 2001–2002. In addition, several lakes in Upstate New York also remained liquid for the first time in at least three decades. In his thirty years of studying freeze-thaw cycles of lakes in New York State, Kenton Stewart had never before seen some of these lakes remain unfrozen for an entire winter. "The majority of the lakes in the state still froze, but a surprising number that developed ice covers in previous winters, had only a partial skim of ice this winter, or did not freeze at all," said Stewart, professor emeritus of biological sciences at the University of Buffalo ("New York Lakes" 2002). In subsequent winters, however, temperatures cooled, and the lakes froze again.

Shortening freezing seasons are not restricted to New York lakes, of course. The waters of Lake Mendota near Madison, Wisconsin, for example, are now frozen about forty fewer days each year than during 1860 (Glick 2004, 32).

According to a report by the Environment News Service, Stewart, who studies the freeze-thaw cycles of more than 250 lakes in New York State, said that "lakes that did not freeze this winter include some that did so during an El Niño year. Those that did freeze did so one to three weeks later than usual.... One surprising thing about the unusually mild winter is that while it was as mild as some of the strong El Niño events that we've seen, it was not associated with an El Niño event in the Pacific Ocean that can have an atmospheric influence. It also was not foreseen by the Climate Prediction Center of the National Oceanographic and Atmospheric Agency" ("New York Lakes" 2002).

Among the New York lakes that failed to freeze during 2001–2002 were Irondequoit Bay in Rochester; Hemlock and Canadice lakes, south of Rochester; Cross Lake located west of Syracuse; Onondaga Lake in Syracuse; Otisco Lake located west of Syracuse; Big Green Lake in Green Lake State Park, east of Syracuse; and Ashokan and other water supply reservoirs north of New York City ("New York Lakes" 2002).

MIGHT HUMAN-INITIATED GLOBAL WARMING END THE ICE AGE CYCLE?

Is it possible that ongoing global warming could delay the onset of the next ice age by thousands of years? Belgian researchers raised this issue in the August 23, 2002, issue of *Science*. "We've shown that the input of greenhouse gas could have an impact on the climate 50,000 years in the future," said Marie-France Loutre of the Universite Catholique de Louvain in Belgium, who researched the question with colleague Andre Berger (Berger and Loutre 2002, 1287; Flam 2002).

Princeton climatologist Jorge Sarmiento said that his own work supports Loutre's assertion that increasing levels of carbon dioxide could linger for thousands of years, long enough to influence the climate of the far future. "The warming will certainly launch us into a new interval in terms of climate, far outside what we've seen before," said Duke University climatologist Tom Crowley. He said it's a big enough influence to cause the cycle of ice ages to "skip a beat" (Flam 2002).

Loutre and Berger estimated that human activity will double the concentration of carbon dioxide in the atmosphere over the next century, raising temperatures as much as 10 degrees F. Still, "It could get much worse," said Crowley. "There's a huge reservoir of coal and if people keep burning it, they could more than quadruple the present carbon dioxide concentrations. I find it hard to believe we will restrain ourselves. It's really rather startling the changes that people will probably see" (Flam 2002). "The silliest thing people could say is: We've got an ice age coming, so why are we worrying about global warming?" Sarmiento said. Whether Loutre and Berger's theory is right or not, "we're going to get a lot of global warming before the ice age kicks in" (2002).

It is possible, over the next several human generations, writes Doug Macdougall in *Frozen Earth: The Once and Future Story of Ice Ages* (2004), that

> warming would be reinforced by the loss of highly reflective ice and snow, and possibly by the decomposition of unstable methane hydrates. The elevated temperatures coupled with the complete loss of continental ice sheets might constitute a threshold-crossing event that would thrust the Earth into a regime from which glaciers could not quickly recover, even with the return of greater [orbital] eccentricity and lower CO_2 levels. Only a few hundred years after Louis Agassiz announced his theory of a global ice age, mankind may inadvertently bring the Pleistocene Ice Age to a premature close, ushering in another long period of ice-free existence for our planet. (p. 244)

REFERENCES: PART III. ICEMELT AROUND THE WORLD

Ainley, D. G., G. Ballard, S. D. Emslie, W. R. Fraser, P. R. Wilson, E. J. Woehler, et al. "Adélie Penguins and Environmental Change." Letter to the Editor. *Science* 300 (April 18, 2003): 429.

"Alaskan Glaciers Retreating." Environment News Service, December 11, 2001. http://ens-news.com/ens/dec2001/2001L-12-11-09.html.

Alley, Richard B. "On Thickening Ice?" *Science* 295 (January 18, 2002): 451–452.

"Antarctic Sea Ice Has Increased." Environment News Service, August 23, 2002. http://ens-news.com/ens/aug2002/2002-08-23-09.asp#anchor5.

Arendt, Anthony A., Keith A. Echelmeyer, William D. Harrison, Craig S. Lingle, and Virginia B. Valentine. "Rapid Wastage of Alaska Glaciers and Their Contribution to Rising Sea Level." *Science* 297 (July 19, 2002): 382–386.

Arthur, Charles. "Snows of Kilimanjaro Will Disappear by 2020, Threatening World-Wide Drought." *London Independent*, October 18, 2002, 7.

Atkinson, Angus, Volker Siegel, Evgeny Pakhomov, and Peter Rothery. "Long-Term Decline in Krill Stock and Increase in Salps within the Southern Ocean." *Nature* 432 (November 4, 2004): 100–103.

Bala, G., K. Caldeira, A. Mirin, and M. Wickett. "Multicentury Changes to the Global Climate and Carbon Cycle: Results from a Coupled Climate and Carbon Cycle Model." *Journal of Climate* (November 1, 2005): 4531–4544.

Baldwin, Mark P., David W. J. Thompson, Emily F. Shuckburgh, Warwick A. Norton, and Nathan P. Gillett. "Weather from the Stratosphere?" *Science* 301 (July 18, 2003): 317–318.

Barbraud, C., and H. Weimerskirch. "Emperor Penguins and Climate Change." *Nature* 411 (May 10, 2001): 183–186.

Bell, Jim. "Nunavut Premier Stands Firm on Global Warming." Environment News Service, August 8, 2002. http://ens-news.com/ens/aug2002/2002-08-09-04.asp.

Berger, Andre, and Marie-France Loutre. "Climate: An Exceptionally Long Interglacial Ahead?" *Science* 297 (August 23, 2002): 1287–1288.

"Billions of People May Suffer Severe Water Shortages as Glaciers Melt: World Wildlife Fund." Agence France Presse, November 27, 2003. (Lexis).

Bowen, Jerry. "Dramatic Climate Change in Alaska." CBS News Transcripts, CBS Morning News, August 29, 2002. (Lexis).

Bowen, Mark. *Thin Ice: Unlocking the Secrets of Climate in the World's Highest Mountains.* New York: Henry Holt, 2005.

Boyd, Robert S. "Earth Warming Could Open Up a Northwest Passage." Knight-Ridder Newspapers, *Pittsburgh Post-Gazette*, November 11, 2002, A-1.

"Breakaway Bergs Disrupt Antarctic Ecosystem." Environmental News Service, May 9, 2002. http://ens-news.com/ens/may2002/2002L-05-09-01.html.

Brown, DeNeen L. "Greenland's Glaciers Crumble: Global Warming Melts Polar Ice Cap into Deadly Icebergs." *Washington Post*, October 13, 2002, A-30.

———. "Hamlet in Canada's North Slowly Erodes: Arctic Community Blames Global Warming as Permafrost Starts to Melt and Shoreline." *Washington Post*, September 13, 2003, A-14.

———. "Waking the Dead, Rousing Taboo: In Northwest Canada, Thawing Permafrost Is Unearthing Ancestral Graves." *Washington Post*, October 17, 2001, A-27.

Brown, Paul. "Geographers' Conference: Ice Field Loss Puts Alpine Rivers at Risk: Global Warming Warning to Europe." *London Guardian*, January 5, 2002, 9.

———. "Global Warming Is Killing Us Too, Say Inuit." *London Guardian*, December 11, 2003, 14.

———. "Melting Permafrost Threatens Alps: Communities Face Devastating Landslides from Unstable Mountain Ranges." *London Guardian*, January 4, 2001, 3.

———. "Scientists Warn of Himalayan Floods: Global Warming Melts Glaciers and Produces Many Unstable Lakes." *London Guardian*, April 17, 2002, 13.

"Brush Fires Collapsing Bear Dens." Canadian Press in *Calgary Sun*, November 2, 2002, 18.

Calamai, Peter. "Global Warming Threatens Reindeer." *Toronto Star*, December 23, 2002, A-23.

Campbell, Duncan. "Greenhouse Melts Alaska's Tribal Ways: As Climate Talks Get Under Way in Bonn Today, Some Americans Are Ruing the Warming Their President Chooses to Ignore." *London Guardian*, July 16, 2001, 11.

Capella, Peter. "Europe's Alps Crumbling: Glaciers Melting in Heatwave." Agence France Presse, August 7, 2003. (Lexis).

Chang, Kenneth. "Arctic Ice Is Melting at Record Level, Scientists Say." *New York Times*, December 8, 2002, A-40.

———. "The Melting (or Freezing) of Antarctica: Deciphering Contradictory Climate Patterns Is Largely a Matter of Ice." *New York Times*, April 2, 2002, F-1.

———. "Ozone Hole Is Now Seen as a Cause for Antarctic Cooling." *New York Times*, May 3, 2002, A-16.

———. "Warming Is Blamed for Antarctica's Weight Gain." *New York Times*, May 20, 2005, A-22.

Chapin, F. S., III, M. Sturm, M. C. Serreze, J. P. McFadden, J. R. Key, A. H. Lloyd, A. D. McGuire, T. S. Rupp, A. H. Lynch, J. P. Schimel, J. Beringer, W. L. Chapman, H. E. Epstein, E. S. Euskirchen, L. D. Hinzman, G. Jia, C.-L. Ping, K. D. Tape, C.D.C. Thompson, D. A. Walker, and J. M. Welker. "Role of Land-Surface Changes in Arctic Summer Warming." *Science* 310 (October 28, 2005): 657–660.

Chengappa, Raj. "The Monsoon: What's Wrong with the Weather?" *India Today*, August 12, 2002, 40.

Chylek, Petr, Jason E. Box, and Glen Lesins. "Global Warming and the Greenland Ice Sheet." *Climatic Change* 63 (2004): 201–221.

Clark, P. U., J. X. Mitrovica, G. A. Milne, and M. E. Tamisiea. "Sea-Level Fingerprinting as a Direct Test for the Source of Global Meltwater Pulse." *Science* 295 (March 29, 2002): 2438–2441.

Clavel, Guy. "Global Warming Makes Polar Bears Sweat." Agence France Presse, November 3, 2002.

Clover, Charles. "Geographers' Conference: Alps May Crumble as Permafrost Melts." *London Telegraph*, January 4, 2001, 12.

Comiso, Josefino C. "A Rapidly Declining Perennial Sea Ice Cover in the Arctic." *Geophysical Research Letters* 29 (20) (October 18, 2002): 1956–1960.

———. "Warming Trends in the Arctic from Clear Sky Satellite Observations." *Journal of Climate* 16 (21) (November 1, 2003): 3498–3510.

Commonwealth Study Conferences. Biography, Ms. Rosemarie Kuptana, Former President, Inuit Tapirisat of Canada. 1998. www.csc-alumni.org/1998/bio/Kuptana.htm.

Connor, Steve. "Meltdown: Arctic Wildlife Is on the Brink of Catastrophe: Polar Bears Could Be Decades from Extinction." *London Independent*, November 11, 2004. (Lexis).

———. "Polar Sea Ice Could Be Gone by the End of the Century." *London Independent*, March 10, 2003, 5.

"Conservation of Arctic Flora and Fauna: Arctic Climate Impact Assessment. An Assessment of Consequences of Climate Variability and Change and the Effects of Increased UV in the Arctic Region: A Draft Implementation Plan." United Nations Environment Programme, October 22, 1999. www.grida.no/caff/acia.htm.

Cook, A. J., A. J. Fox, D. G. Vaughan, and J. G. Ferrigno. "Retreating Glacier Fronts on the Antarctic Peninsula over the Past Half-Century." *Science* 308 (April 22, 2005): 541–544.

Cooke, Robert. "Is Global Warming Making Earth Greener?" *Newsday*, September 11, 2001, C-3.

Cunningham, Dennis. "New Video Documents Climate Change Impacts on [the] High Arctic." International Institute for Sustainable Development, November 9, 2000. www.iisd.org/casl/projects/inuitobs.htm.

Curran, Mark A. J., Tas D. van Ommen, Vin I. Morgan, Katrina L. Phillips, and Anne S. Palmer. "Ice Core Evidence for Antarctic Sea Ice Decline since the 1950s." *Science* 302 (November 14, 2003): 1203–1206.

Dalton, Alastair. "Ice Pack Clue to Climate-Change Effects." *The Scotsman*, October 18, 2001, 7.

Davis, Curt H., Yonghong Li, Joseph R. McConnell, Markus M. Frey, and Edward Hanna. "Snowfall-Driven Growth in East Antarctic Ice Sheet Mitigates Recent Sea-Level Rise." *Science* 308 (June 24, 2005): 1898–1901. Posted online May 19, 2005, www.scienceexpress.org.

de Angelis, Hernán, and Pedro Skvarca. "Glacier Surge after Ice Shelf Collapse." *Science* 299 (March 7, 2003): 1560–1562.

Derbyshire, David. "'Heatwave' in the Antarctic Halves Penguin Colony." *London Daily Telegraph*, May 10, 2001, 6.

Domack, Eugene, Diana Duran, Amy Leventer, Scott Ishman, Sarah Doane, Scott McCallum, et al. "Stability of the Larsen B Ice Shelf on the Antarctic Peninsula during the Holocene Epoch." *Nature* 436 (August 4, 2005): 681–685.

Egan, Timothy. "Alaska, No Longer So Frigid, Starts to Crack, Burn, and Sag." *New York Times*, June 16, 2002, A-1.

———. "On Hot Trail of Tiny Killer in Alaska." *New York Times*, June 25, 2002, F-1.

———. "The Race to Alaska before It Melts." *New York Times*, June 26, 2005, Travel Section.

Elliott, Valerie. "Polar Bears Surviving on Thin Ice." *London Times*, October 30, 2003. (Lexis).

English, Philip. "Ross Island Penguins Struggling." *New Zealand Herald*, January 9, 2003. (Lexis).

Erickson, Jim. "Glaciers Doff Their Ice Caps, and as Frozen Fields Melt, Anthropological Riches Are Revealed." *Rocky Mountain News*, August 22, 2002, 6-A.

———. "Going, Going, Gone? Front-Range Glaciers Declining: Researchers Point to a Warming World." *Rocky Mountain News*, October 26, 2004, 5-A.

"Expert Fears Warming Will Doom Bears." Canadian Press in *Victoria Times-Colonist*, January 5, 2003, C-8.

Finch, Gavin. "Falklands Penguins Dying in Thousands." *London Independent*, June 19, 2002, 12.

Flam, Faye. "It's Hot Now, but Scientists Predict There's an Ice Age Coming." *Philadelphia Inquirer*, August 23, 2002. (Lexis).

Foley, Jonathan A. "Tipping Points in the Tundra." *Science* 310 (October 28, 2005): 627–628.

Forero, Juan. "As Andean Glaciers Shrink, Water Worries Grow." *New York Times*, November 24, 2002, A-3.

Fowler, C., W. J. Emery, and J. Maslanik. "Satellite-Derived Evolution of Arctic Sea Ice Age: October 1978 to March 2003." *Geoscience and Remote Sensing Letters* 1 (2) (2004): 71–74.

Frey, Darcy. "George Divoky's Planet." *New York Times Sunday Magazine*, January 6, 2002, 26–30.

Ganopolski, Andrey, and Stefan Rahmstorf. "Rapid Changes of Glacial Climate Simulated in a Coupled Climate Model." *Nature* 409 (January 11, 2001): 153–158.

Gasse, Françoise. "Kilimanjaro's Secrets Revealed." *Science* 298 (October 18, 2002): 548–549.

George, Jane. "Global Warming Threatens Nunavut's National Parks." *Nunatsiaq News*, May 19, 2000. www.nunatsiaq.com/archives/nunavut000531/nvt20519_18.html.

Gillett, Nathan P., and David W. J. Thompson. "Simulation of Recent Southern Hemisphere Climate Change." *Science* 302 (October 10, 2003): 273–275.

"Glacial Retreat Seen Worldwide." Environment News Service, May 30, 2002. http://ens-news.com/ens/may2002/2002-05-30-09.asp#anchor2; or www.gsfc.nasa.gov/topstory/20020530glaciers.html.

Glick, Daniel. "The Heat Is On: Geosigns." *National Geographic*, September 2004, 12–33.

"Global Warming and Freak Winds Combine to Allow Explorers through Northeast Passage." *London Independent*, October 11, 2002, 14.

"Global Warming Makes China's Glaciers Shrink by Equivalent of Yellow River." Agence France Presse, August 23, 2004. (Lexis).

"Global Warming Troubles Qinghai-Tibet Railway Construction." Xinhua (Chinese News Agency), April 30, 2003. (Lexis).

Goldman, Erica. "Even in the High Arctic, Nothing Is Permanent." *Science* 297 (August 30, 2002): 1493–1494.

Gregory, Jonathan M., Philippe Huybrechts, and Sarah C. B. Raper. "Threatened Loss of the Greenland Ice Sheet." *Nature* 428 (April 8, 2004): 616.

Hall, Alan. "Postcards a Tell-Tale for Icy Retreat." *The Scotsman*, August 3, 2002, 8.

Hall, Alex, and Ronald J. Stouffer. "An Abrupt Climate Event in a Coupled Ocean-Atmosphere Simulation without External Forcing." *Nature* 409 (January 11, 2001): 171–174.

Harvey, Fiona. "Arctic May Have No Ice in Summer by 2070, Warns Climate Change Report." *London Financial Times*, November 2, 2004, 1.

Henderson, Mark. "Antarctica Defies Global Warming." *London Times*, January 14, 2002.

———. "Southern Krill Decline Threatens Whales, Seals." *London Times* in *Calgary Herald*, November 4, 2004, A-11.

Hodge, Amanda. "Patagonia's Big Melt 'Sign of Global Warming.'" *The (Sydney) Australian*, February 12, 2004, 8.

Howat, I. M., I. Joughin, S. Tulaczyk, and S. Gogineni. "Rapid Retreat and Acceleration of Helheim Glacier, East Greenland." *Geophysical Research Letters* 32 (2005). L22502, doi:10.1029/2005GL024737.

"Hudson Bay Ice-Free by 2050, Scientists Say." Canadian Broadcasting Corporation, March 14, 2001. http://north.cbc.ca/cgi-bin/templates/view.cgi?/news/2001/03/14/14hudsonice.

Human, Katy. "Disappearing Arctic Ice Chills Scientists: A University of Colorado Expert on Ice Worries That the Massive Melting Will Trigger Dramatic Changes in the World's Weather." *Denver Post*, October 5, 2004, B-2.

"Ice a Scarce Commodity on Arctic Rinks: Global Warming Blamed for Shortened Hockey Season." *Financial Post (Canada)*, January 7, 2003, A-3.

Jacobs, S. S., C. F. Giulivi, and P. A. Mele. "Freshening of the Ross Sea during the Late 20th Century." *Science* 297 (July 19, 2002): 386–389.

Johannessen, Ola M., Kirill Khvorostovsky, Martin W. Miles, and Leonid P. Bobylev. "Recent Ice-Sheet Growth in the Interior of Greenland." *Science* 310 (November 11, 2005): 1013–1016.

Johansen, Bruce E. "Arctic Heat Wave." *The Progressive*, October 2001, 18–20.

Joughin, Ian, and Slawek Tulaczyk. "Positive Mass Balance of the Ross Ice Streams, West Antarctica." *Science* 295 (January 18, 2002): 476–480.

Kaiser, Jocelyn. "Glaciology: Warmer Ocean Could Threaten Antarctic Ice Shelves." *Science* 302 (October 31, 2003): 759.

Karoly, David J. "Ozone and Climate Change." *Science* 302 (October 10, 2003): 236–237.

Kaser, Georg, Douglas R. Hardy, Thomas M. A. Org, Raymond S. Bradley, and Tharsis M. Hyera. "Modern Glacier Retreat on Kilimanjaro as Evidence of Climate Change: Observations and Facts." *International Journal of Climatology* 24 (2004): 329–339.

Kerr, Richard A. "A Single Climate Mover for Antarctica." *Science* 296 (May 3, 2002): 825–826.

———. "A Warmer Arctic Means Change for All." *Science* 297 (August 30, 2002): 1490–1493.

Konviser, Bruce I. "Glacier Lake Puts Global Warming on the Map." *Boston Globe*, July 16, 2002, C-1.

Krajick, Kevin. "Arctic Life, on Thin Ice." *Science* 291 (January 19, 2001): 424–425.

———. "Ice Man: Lonnie Thompson Scales the Peaks for Science." *Science* 298 (October 18, 2002): 518–522.

———. "Tracing Icebergs for Clues to Climate Change." *Science* 292 (June 22, 2001): 2244–2245.

Krauss, Clifford, Steven Lee Myers, Andrew C. Revkin, and Simon Romero. "As Polar Ice Turns to Water, Dreams of Treasure Abound." *New York Times*, October 10, 2005. www.nytimes.com/2005/10/10/science/10arctic .html?ei=5094&en=64e93c8fc877d5f2&hp=&ex=1129003200&partner= homepage&pagewanted=print.

Kristal-Schroder, Carrie. "British Adventurer's Polar Trek Foiled by Balmy Arctic." *Ottawa Citizen*, May 17, 2004, A-1.

Kristof, Nicholas D. "Baked Alaska on the Menu*?*" *New York Times* in *Alameda (California) Times-Star*, September 14, 2003. (Lexis).

"Late Snow Smoothes the Way for Iditarod Sled-Dog Race." Reuters in *Ottawa Citizen*, March 2, 2001, B-8.

Lawrence, David M., and Andrew G. Slater. "A Projection of Severe Near-surface Permafrost Degradation during the 21st Century." *Geophysical Research Letters* 32 (2005). L24401, doi:10.1029/2005GL025080.

Laxon, Seymour, Neil Peacock, and Doug Smith. "High Interannual Variability of Sea-Ice Thickness in the Arctic Region." *Nature* 425 (October 30, 2003): 947–950.

Lean, Geoffrey. "Antarctic Becomes Too Hot for the Penguins: Decline of 'Dinner Jacket' Species Is a Warning to the World." *London Independent*, February 3, 2002, 9.

———. "The Big Thaw: Global Disaster Will Follow If the Ice Cap on Greenland Melts; Now Scientists Say It is Vanishing Far Faster than Even They

Expected." *London Independent*, November 20, 2005. www.commondreams .org/headlines05/1120-03.htm.

Leggett, Jeremy. *The Carbon War: Global Warming and the End of the Oil Era*. New York: Routledge, 2001.

Linden, Eugene. "The Big Meltdown: As the Temperature Rises in the Arctic, It Sends a Chill around the Planet." *Time* 156, September 4, 2000. www.time .com/time/magazine/articles/0,3266,53418,00.html.

Loeb, V. "Effects of Sea-Ice Extent and Krill or Salp Dominance on the Antarctic Food Web." *Nature* 387 (1997): 897–900.

Lynas, Mark. *High Tide: News from a Warming World*. London: Flamingo, 2004.

———. *High Tide: The Truth about Our Climate Crisis*. New York: Picador/St. Martins, 2004.

———. "Meltdown: Alaska Is a Huge Oil Producer and Has Become Rich on the Proceeds. But It Has Suffered the Consequences: Global Warming, Faster and More Terrifyingly Than Anyone Could Have Predicted." *London Guardian* (Weekend Magazine), February 14, 2004, 22.

———. "Vanishing Worlds: A Family Snap[shot] of a Peruvian Glacier Sent Mark Lynas on a Journey of Discovery: With the Ravages of Global Warming, Would It Still Exist 20 Years Later?" *London Guardian*, March 31, 2004, 12.

Macdougall, Doug. *Frozen Earth: The Once and Future Story of Ice Ages*. Berkeley: University of California Press, 2004.

"Major Temperature Rise Recorded in Arctic This Year: German Scientists." Agence France Presse, August 27, 2004. (Lexis).

McFarling, Usha Lee. "Glacial Melting Takes Human Toll: Avalanche in Russia and Other Disasters Show That Global Warming Is Beginning to Affect Areas Much Closer to Home." *Los Angeles Times*, September 25, 2002, A-4.

———. "Shrinking Ice Cap Worries Scientists." *Los Angeles Times* in *Edmonton Journal*, December 8, 2002. www.canada.com/regina/story.asp?id= {54910725-535A-4B0E-9A7E-FD7176D9C392}.

McKie, Robin. "Decades of Devastation Ahead as Global Warming Melts the Alps: A Mountain of Trouble as Matterhorn Is Rocked by Avalanches." *London Observer*, July 20, 2003, 18.

McLean, Jim. "Icebergs Fuel Fears on Climate." *Glasgow Herald*, December 11, 2000, 8.

Meier, Mark F., and Mark B. Dyurgerov. "Sea-Level Changes: How Alaska Affects the World." *Science* 297 (July 19, 2002): 350–351.

"Melting of Ice at the North Pole to be Surveyed." Inuit Circumpolar Conference Memo, January 8, 2002. From the files of Sheila Watt-Cloutier and Christian Schultz-Lorentzen.

"Melting Planet: Species are Dying Out Faster Than We Have Dared Recognize, Scientists Will Warn This Week." *London Independent*, October 2, 2005. http://news.independent.co.uk/world/environment/article316604.ece.

Meuvret, Odile. "Global Warming Could Turn Siberia into Disaster Zone: Expert." Agence France Presse, October 2, 2003.

Monastersky, Richard. "The Long Goodbye: Alaska's Glaciers Appear to Be Disappearing before Our Eyes: Are They a Sign of Things to Come?" *New Scientist* (April 14, 2001): 30–32.

"Monster Iceberg Heads into Antarctic Waters." Agence France Presse, October 22, 2002. (Lexis).

Montaigne, Fen. "The Heat Is On: Ecosigns." *National Geographic*, September 2004, 34–55.

Moran, Tom. "Scientists Trace Global Warming in [Alaska's] Interior." Associated Press, Alaska State Wire, July 10, 2003. (Lexis).

"More Evidence Found of Warming in the Alaskan Arctic." Associated Press in *Omaha World-Herald*, June 3, 2001, 15-A.

"More Than 80 Per Cent of Spain's Pyrenean Glaciers Melted Last Century." Agence France Presse, September 29, 2004. (Lexis).

Morton, Oliver. "The Tarps of Kilimanjaro." *New York Times*, November 17, 2003. www.nytimes.com/2003/11/17/opinion/17MORT.html.

Murphy, Kim. "Front-Row Exposure to Global Warming: Engineers say Alaskan Village Could Be Lost as Sea Encroaches." *Los Angeles Times*, July 8, 2001, A-1.

Naftz, David L., David D. Susong, Paul F. Schuster, L. DeWayne Cecil, Michael D. Dettinger, Robert L. Michel, et al. "Ice Core Evidence of Rapid Air Temperature Increases since 1960 in the Alpine Areas of the Wind River Range, Wyoming, United States." *Journal of Geophysical Research* 107 (July 9, 2002): 4171–4187. www.agu.org/pubs/crossref/2002/2001JD000621.shtml.

Naish, T. R., K. J. Woolfe, P. J. Barnett, G. S. Wilson, C. Atkins, S. M. Bohaty, et al. "Orbitally Induced Oscillations in the East Antarctic Ice Sheet at the Oligocene/Miocene Boundary." *Nature* 413 (October 18, 2001): 719–723.

Nash, J. Madeleine. *El Niño: Unlocking the Secrets of the Master Weather-Maker*. New York: Warner Books, 2002.

Nesmith, Jeff. "Antarctic Glacier Melt Increases Dramatically." *Atlanta Journal-Constitution*, September 22, 2004, 9-A.

"New Research Reveals 50-Year Sustained Antarctic Ice Decline." Agence France Presse, November 14, 2003. (Lexis).

"New York Lakes Fail to Freeze." Environment News Service, March 21, 2002. http://ens-news.com/ens/mar2002/2002L-03-21-09.html#anchor1.

Nickerson, Colin. "An Early Melting Hurts Seals, Hunters in Canada." *Boston Globe*, April 1, 2002, A-1.

"North Pole Mussels Point to Warming, Scientists Say." Reuters in *Canada National Post*, September 18, 2004, A-16.

"100 Still Missing in Russian Avalanche." *Los Angeles Times*, September 24, 2002, A-5.

Osterkamp, T., and V. Romanovsky. "Permafrost Monitoring and Detection of Climate Change: Comments." *Permafrost and Periglacial Processes* 9 (1998): 87–89.

Overpeck, J. T., M. Strum, J. A. Francis, D. K. Perovich, M. C. Serreze, R. Benner, E. C. Carmack, F. S. Chapin III, S. C. Gerlach, L. C. Hamilton, L. D. Hinzman, M. Holland, H. P. Huntington, J. R. Key, A. H. Lloyd, G. M. MacDonald, J. McFadden, D. Noone, T. D. Prowse, P. Schlosser, and C. Vorosmarty. "Artic System on Trajectory to New, Seasonally Ice-Free State." *EOS: Transactions of the American Geophysical Union* 86 (34) (August 23, 2005): 309, 312.

Paillard, Didler. "Glacial Hiccups." *Nature* 409 (January 11, 2001): 147–148.

Paraskevas, Joe. "Glaciers in the Canadian Rockies Shrinking to Their Lowest Level in 10,000 Years." *Canada National Post*, December 4, 2003, A-8.

Paterson, W.S.B., and N. Reeh. "Thinning of the Ice Sheet in Northwest Greenland over the Past Forty Years." *Nature* 414 (November 1, 2001): 60–62.

Payne, A. J., A. Vieli, A. P. Shepherd, D. J. Wingham, and E. Rignot. "Recent Dramatic Thinning of Largest West Antarctic Ice Stream Triggered by Oceans." *Geophysical Research Letters* 31 (23) (December 9, 2004). L23401, doi:10.1029/2004GL021284.

Pegg, J. R. "The Earth Is Melting, Arctic Native Leader Warns." Environment News Service, September 16, 2004.

Perlman, David. "Shrinking Glaciers Evidence of Global Warming: Differences Seen by Looking at Photos from 100 Years Ago." *San Francisco Chronicle*, December 17, 2004, A-18.

Petit, Charles W. "Arctic Thaw." *U.S. News and World Report*, November 8, 2004, 66–69.

Phelan, Amanda. "Turning Up the Heat." *Sydney Sunday Telegraph*, August 18, 2002, 47.

Pianin, Eric. "Study Fuels Worry over Glacial Melting: Research Shows Alaskan Ice Mass Vanishing at Twice Rate Previously Estimated." *Washington Post*, July 19, 2002, A-14.

Putkonen, J. K., and G. Roe. "Rain-on-Snow Events Impact Soil Temperatures and Affect Ungulate Survival." *Geophysical Research Letters* 30 (4) (2003): 1188–1192.

Quayle, Wendy C., Lloyd S. Peck, Helen Peat, J. C. Ellis-Evans, and P. Richard Harrigan. "Extreme Responses to Climate Change in Antarctic Lakes." *Science* 295 (January 25, 2002): 645.

Radford, Tim. "Antarctic Ice Cap Is Getting Thinner: Scientists' Worries That the South Polar Ice Sheet Is Melting May Be Confirmed by the Dramatic Retreat of the Region's Biggest Glacier." *London Guardian*, February 2, 2001, 9.

———. "85 Per Cent of Alaskan Glaciers Melting at 'Incredible Rate.'" *London Guardian*, July 19, 2002, 9.

Ralston, Greg. "Study Admits Arctic Danger." *Yukon News*, November 15, 1996. http://yukonweb.com/community/yukon-news/1996/nov15.htmld/#study.

"Rare Sighting of Wasp North of Arctic Circle Puzzles Residents." Canadian Broadcasting Corporation, September 9, 2004. www.cbc.ca/story/science/national/2004/09/09/wasp040909.html.

"Recent Warming of Arctic May Affect World-Wide Climate." National Aeronautics and Space Administration Press Release, October 23, 2003. www.gsfc.nasa.gov/topstory/2003/1023esuice.html.

Reid, K., and J. P. Croxall. "Environmental Response of Upper Trophic-Level Predators Reveals a System Change in an Antarctic Marine Ecosystem." *Proceedings of the Royal Society of London* B268 (2001): 377–384.

Remington, Robert. "Goodbye to Glaciers: Thanks to Global Warming, Mountains—the World's Water Towers—Are Losing Their Ice. As It Disappears, so Does an Irreplaceable Source of Water." *Financial Post (Canada)*, September 6, 2002, A-19.

Rergaldo, Antonio. "The Ukukus Wonder Why a Sacred Glacier Melts in Peru's Andes." *Wall Street Journal*, June 17, 2005, A-1, A-10.

Revkin, Andrew C. "Antarctic Glaciers Quicken Pace to Sea: Warming Is Cited." *New York Times*, September 24, 2004, A-24.

———. "A Chilling Effect on the Great Global Melt." *New York Times*, January 18, 2002, A-17.

———. "Climate Debate Gets Its Icon: Mt. Kilimanjaro." *New York Times,* March 23, 2004. NYTimes.com.

———. "Forecast for a Warmer World: Deluge and Drought." *New York Times*, August 28, 2002, A-10.

———. "An Icy Riddle as Big as Greenland." *New York Times*, June 8, 2004. www.nytimes.com/2004/06/08/science/earth/08gree.html.

———. "A Message in Eroding Glacial Ice: Humans Are Turning Up the Heat." *New York Times*, February 19, 2001, A-1.

———. "New Climate Model Highlights Arctic's Vulnerability." *New York Times*, October 31, 2005. www.nytimes.com/2005/10/31/science/earth/01warm_web.html.

————. "Study of Antarctic Points to Rising Sea Levels." *New York Times*, March 7, 2003, A-8.

Rignot, Eric, G. Casassa; P. Gogineni, W. Krabill, A. Rivera, and R. Thomas. "Accelerated Ice Discharge from the Antarctic Peninsula Following the Collapse of Larsen B Ice Shelf." *Geophysical Research Letters* 31 (18) (September 22, 2004). doi:10.1029/2004GL020697.

————, and Stanley S. Jacobs. "Rapid Bottom Melting Widespread Near Antarctic Ice Sheet Grounding Lines." *Science* 296 (June 14, 2002): 2020–2023.

————, Andrès Rivera, and Gino Casassa. "Contribution of the Patagonia Icefields of South America to Sea Level Rise." *Science* 302 (October 17, 2003): 434–437.

————, and Robert H. Thomas. "Mass Balance of Polar Ice Sheets." *Science* 297 (August 30, 2002): 1502–1506.   .

Rigor, I. G., and J. M. Wallace. "Variations in the Age of Arctic Sea-Ice and Summer Sea-Ice Extent." *Geophysical Research Letters* 31 (2004). doi:10.1029/2004GL019492.

Robson, Seth. "Glaciers Melting." *Christchurch (New Zealand) Press*, February 26, 2003, 13.

Roe, Nicholas. "Show Me a Home Where the Reindeer Roam." *London Times*, November 10, 2001. (Lexis).

Rohter, Larry. "Antarctica, Warming, Looks Ever More Vulnerable." *New York Times*, January 25, 2005. www.nytimes.com/2005/01/25/science/earth/25ice.html.

Rosen, Yereth. "Alaska's Not-So-Permanent Frost." *Christian Science Monitor*, October 7, 2003, 1.

Russell, Sabin. "Glaciers on Thin Ice: Expert Says Melting to Be Faster Than Expected." *San Francisco Chronicle*, February 17, 2002, A-4.

Sabadini, Roberto. "Ice Sheet Collapse and Sea-Level Change." *Science* 295 (March 29, 2002): 2376–2377.

Sample, Ian. "Warming Hits Tipping Point: Climate Change Alarm as Siberian Permafrost Melts for First Time since Ice Age." *Manchester Guardian Weekly*, August 18, 2005, 1. www.guardian.co.uk/guardianweekly/story/0,12674,1550685,00.html.

Scambos, T. A., J. A. Bohlander, C. A. Shuman, and P. Skvarca. "Glacier Acceleration and Thinning after Ice Shelf Collapse in the Larsen B Embayment, Antarctica." *Geophysical Research Letters* 31 (18) (September 22, 2004). doi:10.1029/2004GL020670.

Schiermeier, Quirin. "A Rising Tide: The Ice Covering Greenland Holds Enough Water to Raise the Oceans Six Metres—and It's Starting to Melt." *Nature* 428 (March 11, 2004): 114–115.

Schmid, Randolph E. "Panel Sees Growing Threat in Melting Arctic." Associated Press, August 23, 2005. (Lexis).

"Scientists Feel Need for Urgency in Arctic Climate Research." Canadian Broadcasting Corporation, April 23, 2001. http://cbc.ca/cgibin/templates/view.cgi?/news/2001/04/23/nunavut_stu_01 0423.

Semenov, Vladimir A., and Lennart Bengtsson. *Modes of Wintertime Arctic Temperature Variability*. Hamburg, Germany: Max Planck Institut fur Meteorologie, 2003.

Semmens, Grady. "Ecologists See Disaster in Dwindling Water Supply." *Calgary Herald*, November 27, 2003, A-14.

Serreze, M. C., J. A. Maslanik, T. A. Scambos, F. Fetterer, J. Stroeve, K. Knowles, et al. "A Record Minimum Arctic Sea Ice Extent and Area in 2002." *Geophysical Research Letters* 30 (2003). doi:10.1029/2002GL016406.

Shepherd, Andrew, Duncan J. Wingham, Justin A. D. Mansley, and Hugh F. J. Corr. "Inland Thinning of Pine Island Glacier, West Antarctica." *Science* 291 (February 2, 2001): 862–864.

———, Duncan Wingham, Tony Payne, and Pedro Skvarca. "Larsen Ice Shelf Has Progressively Thinned." *Science* 302 (October 31, 2003): 856–859.

———, D. Wingham, and E. Rignot. "Warm Ocean Is Eroding West Antarctic Ice Sheet." *Geophysical Research Letters* 31 (23) (December 9, 2004). L23402, doi:10.1029/2004GL021106.

Shindell, Drew T. "Climate Change: Whither Arctic Climate?" *Science* 299 (January 10, 2003): 215–216.

———, and G. A. Schmidt. "Southern Hemisphere Climate Response to Ozone Changes and Greenhouse Gas Increases." *Geophysical Research Letters* 31 (2004). L18209, doi:10.1029/2004GL020724.

Small, Jason. "Caskets, Bodies Surface during Frost Heaves in Yukon." *Whitehorse Star* in *Ottawa Citizen*, July 21, 2001, G-8.

Smith, L. C., Y. Sheng, G. M. MacDonald, and L. D. Hinzman. "Disappearing Arctic Lakes." *Science* 308 (June 3, 2005): 1429.

Spears, Tom. "Antarctica Rides Global 'Heatwave': Continent's Warm Coast Causes Concern." *Ottawa Citizen*, August 8, 2001, A-1.

Stirling, Ian, and Andrew Derocher. "Possible Impacts of Climatic Warming on Polar Bears." *Arctic* 46 (3) (September 1993): 240–245.

Stokstad, Eric. "Defrosting the Carbon Freezer of the North." *Science* 304 (June 11, 2004): 1618–1620.

Stroeve, J. C., M. C. Serreze, F. Fetterer, T. Arbetter, W. Meier, J. Maslanik, et al. "Tracking the Arctic's Shrinking Ice Cover: Another Extreme September Minimum in 2004." *Geophysical Research Letters* 32 (4) (February 25, 2005). L04501. http://dx.doi.org/10.1029/2004GL021810.

Sturm, Matthew, Josh Schimel, Gary Michaelson, Jeffery M. Welker, Steven F. Oberbauer, Glen E. Liston, et al. "Winter Biological Processes Could Help Convert Arctic Tundra to Shrubland." *BioScience* 55 (1) (January 2005): 17–26.

Sturm, M., C. Racine, and K. Tape. "Climate Change: Increasing Shrub Abundance in the Arctic." *Nature* 411 (May 31, 2001): 546–548.

"Thin Polar Bears Called Sign of Global Warming." Environment News Service, May 16, 2002. http://ens-news.com/ens/may2002/2002L-05-16-07.html.

Thomas, R., E. Rignot, G. Casassa, P. Kanagaratnam, C. Acuòa, T. Akins, et al. "Accelerated Sea-Level Rise from West Antarctica." *Science* 306 (October 8, 2004): 255–258.

Thompson, David W. J., and Susan Solomon. "Interpretation of Recent Southern Hemisphere Climate Change." *Science* 296 (May 3, 2002): 895–899.

Thompson, Lonnie G., Ellen Mosley-Thompson, Mary E. Davis, Keith A. Henderson, Henry H. Brecher, Victor S. Zagorodnov, et al. "Kilimanjaro Ice Core Records: Evidence of Holocene Climate Change in Tropical Africa." *Science* 298 (October 18, 2002): 589–593.

"The Threat of Climate Change to Arctic Wildlife." Greenpeace, May 1998. www.greenpeace.org/~climate/arctic99/reports/wildlife.html.

"Time to Act on Nunavut Climate Change." Canadian Broadcasting Corporation, March 16, 2001. http://north.cbc.ca/cgi-bin/templates/view.cgi?/news/2001/03/16/16nunmoose.

Toner, Mike. "Arctic Ice Thins Dramatically, NASA Satellite Images Show." *Atlanta Journal and Constitution*, October 24, 2003, 1-A.

———. "Huge Ice Chunk Breaks off Antarctica." *Atlanta Journal and Constitution*, March 20, 2002, A-1.

———. "Meltdown in Montana: Scientists Fear Park's Glaciers May Disappear within 30 Years." *Atlanta Journal and Constitution*, June 30, 2002, 4-A.

Toniazzo, T., J. M. Gregory, and P. Huybrechts. "Climatic Impact of a Greenland Deglaciation and Its Possible Irreversibility." *Journal of Climate* 17 (1) (January 1, 2004): 21–33.

Vaughan, David G., G. J. Marshall, W. M. Connolley, C. L. Parkinson, R. Mulvaney, D. A. Hodgson, et al. "Recent Rapid Regional Climate Warming on the Antarctic Peninsula." *Climatic Change* 60 (3) (October 2003): 243–274.

Verrengia, Joseph B. "In Alaska, an Ancestral Island Home Falls Victim to Global Warming: 'We Have No Room Left.'" Associated Press, September 9, 2002. (Lexis).

Vidal, John. "Antarctica Sends Warning of the Effects of Global Warming: Scientists Stunned as Ice Shelf Falls Apart in a Month." *London Guardian*, March 20, 2002, 3.

————. "Mountain Cultures in Grave Danger Says U.N.: Agriculture, Climate and Warfare Pose Dire Threat to Highland Regions around the World." *London Guardian*, October 24, 2002, 7.

Von Radowitz, John. "Antarctic Wildlife 'at Risk from Global Warming.' " Press Association News, September 9, 2002. (Lexis).

Watt-Cloutier, Sheila. Speech to Conference of Parties to the United Nations Framework Convention on Climate Change. Milan, Italy, December 10, 2003. www.inuitcircumpolar.com/section.php?ID=8&Lang=En&Nav=Section.

Weber, Bob. "Arctic Sea Ice Isn't Melting, Just Drifting Away: New Study." *Montreal Gazette*, April 25, 2001, A-8.

Whipple, Dan. "Climate: The Arctic Goes Bush." United Press International, January 10, 2005. (Lexis).

Whitfield, John. "Alaska's Climate: Too Hot to Handle." *Nature* 425 (September 25, 2003): 338–339.

Wilkin, Dwayne. "Global Warming Poses Big Threats to Canada's Arctic." *Nunatsiaq News*, November 21, 1997. www.nunatsiaq.com/archives/back issues/71121.html#6.

————. "A Team of Glacial Ice Experts Say Mother Nature's Thermostat Has Kept the Eastern Arctic at about the Same Temperature since 1960." *Nunatsiaq News*, May 30, 1997. www.nunanet.com/~nunat/week/70530.html#7.

Williams, Frances. "Everest Hit by Effects of Global Warming." *London Financial Times*, June 6, 2002, 2.

Wilson, Scott. "Warming Shrinks Peruvian Glaciers: Retreat of Andean Snow Caps Threatens Future for Valleys." *Washington Post*, July 9, 2001, A-1.

Wines, Michael. "Rising Star Lost in Russia's Latest Disaster." *New York Times*, September 24, 2002, A-11.

Wohlforth, Charles. *The Whale and the Supercompter: On the Northern Front of Climate Change*. New York: North Point Press/Farrar, Strauss and Giroux, 2004.

Wolff, Eric W. "Whither Antarctic Sea Ice?" *Science* 302 (November 14, 2003): 1164.

Woods, Michael. "Atop World, Scientists Ask Why Ice Cap Is Melting." *Pittsburgh Post-Gazette*, April 26, 1999. www.post-gazette.com/healthscience/19990423ice1.asp.

Yoon, Carol Kaesuk. "Penguins in Trouble Worldwide." *New York Times*, June 26, 2001, F-1.

Zwally, H. Jay, Waleed Abdalati, Tom Herring, Kristine Larson, Jack Saba, and Konrad Steffen. "Surface Melt-Induced Acceleration of Greenland Ice-Sheet Flow." *Science* 297 (July 12, 2002): 218–222.

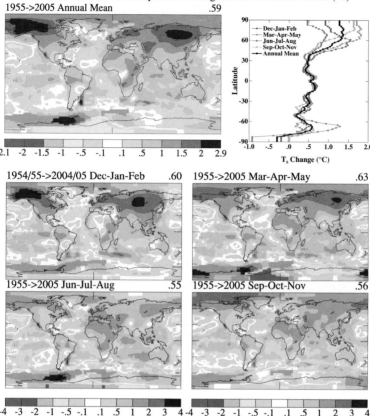

Last 50 Years Surface Temperature Change Based on Linear Trends (°C)

1955->2005 Annual Mean .59

1954/55->2004/05 Dec-Jan-Feb .60 1955->2005 Mar-Apr-May .63

1955->2005 Jun-Jul-Aug .55 1955->2005 Sep-Oct-Nov .56

Temperatures worldwide, 1955–2005, compared to average. Cumulative at top; followed by winter, middle, left; spring, middle, right; summer, bottom, left; autumn, bottom, right. The graph at the top right plots the change by latitude, illustrating greater warming in polar areas. Scale in degrees C. NASA, Goddard Institute for Space Studies.

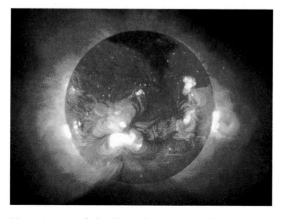

X-ray image of the Sun. Courtesy of Getty Images/ PhotoDisc.

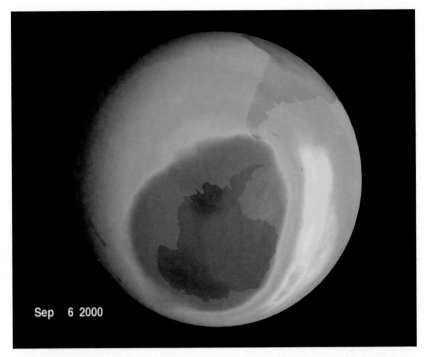

Sep 6 2000

Stratospheric ozone depletion over Antarctica at its greatest extent, September 6, 2000. Courtesy of NASA.

Lake Michigan. Courtesy of Getty Images/Harcourt.

The Edith's checkerspot. Courtesy of Adam Winer.

Tokyo at night. Courtesy of Getty Images/PhotoDisc.

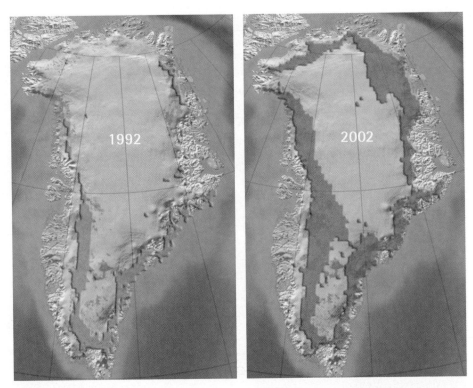

Summer snowmelt on Greenland's ice cap, 1992 and 2002. Courtesy of Clifford Grabhorn.

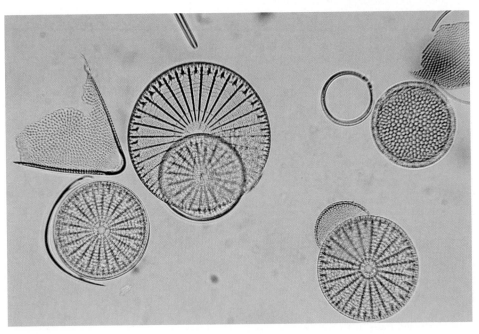

Diatoms (Phytoplankton). Courtesy of Getty Images/PhotoDisc.

People walk in the morning through smoke from a burning forest September 5, 2002 in Moscow, as forests and peat bogs burned near the city. © Oleg Nikishin/Getty Images.

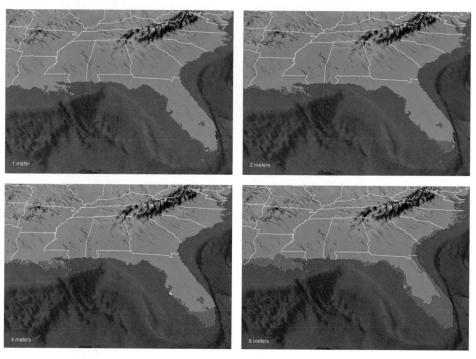

Prospective land losses in Florida with a sea-level rise of one to four meters. Courtesy of NOAA/ Geophysical Fluid Dynamics Laboratory.

Damage caused to forests by spruce bark beetle infestation near Homer, on the Kenai Peninsula, Alaska. © Daniel Beltra/AFP/Getty Images.

Aerial view of the Great Barrier Reef. Courtesy of Corbis.

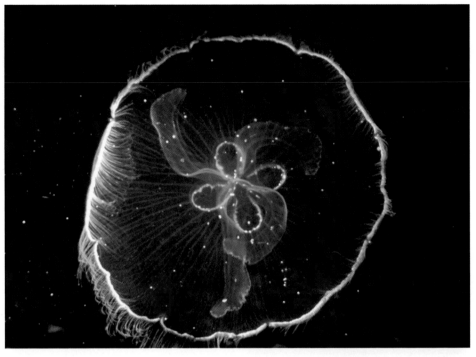

A moon jellyfish. Courtesy of the National Oceanic and Atmospheric Administration Photo Library.

2005 Surface Temperature Anomaly (°C)

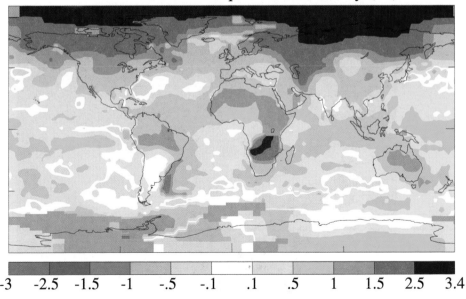

| -3 | -2.5 | -1.5 | -1 | -.5 | -.1 | .1 | .5 | 1 | 1.5 | 2.5 | 3.4 |

Temperatures worldwide for 2005, compared to average. Scale in degrees C. NASA, Goddard Institute for Space Studies.

Monteverde golden toad. Courtesy of Charles
H. Smith/U.S. Fish and Wildlife Service.

Tufted puffin. Courtesy of the National Oceanic and
Atmospheric Administration Photo Library.

Hummingbird hawk moth. Photo by Steven Foster.

IV WARMING SEAS

INTRODUCTION

On a practical level, rising seas provoked by melting ice and thermal expansion of seawater will be the most notable challenge related to global warming (ranging from inconvenience to disaster) for many people around the world. Human beings have an affinity for the open sea. Thus, many major population centers have been built within a mere meter or two of mean sea level. From Bombay to London to New York City, many millions of people will find warming seawater lapping at their heels during coming years. Sea levels have been rising very slowly for a century or more, and the pace will increase in coming years.

Less visible, but just as profound, warming seas could affect patterns of oceanic flow around the world (the Thermohaline Circulation) that replenish the world's oceans with oxygen, leading to possible extinction of several sea creatures. Additionally, a large proportion of the planet's coral reefs are already suffering some degree of heat stress from warming waters. Additional warming could wipe out many of these "rainforests of the sea." Many fisheries will change as cold-water fish move north or become extinct. Sea species usually identified with the tropics already are wriggling into heretofore cooler waters in places such as Britain and the U.S. Pacific Northwest. At the base of the food chain, phytoplankton stocks have been reduced significantly by oceanic warming in some areas. The coming century promises more of the same.

A team of oceanographers using a combination of observations and climate models had determined by 2005 that the penetration of human-induced warming in the world's oceans was unmistakable. Furthermore, the same team reported that roughly 84 percent of the extra heat that the entire Earth was absorbing had impacted the oceans during the previous forty years (Barnett et al. 2005, 284). Their study concluded that "little doubt" exists "that there is a human-induced signal in the environment." The record also leaves little doubt that the oceans will continue to warm. "How to respond to the serious problems posed by these predictions is a question that society must decide," they wrote in *Science* (p. 287).

As scientists debate the speed with which the oceans are warming, anecdotal evidence abounds. During the summer of 2002, ocean water was warm enough for swimming along much of Maine's coast, a novelty. A local account said that "hot days and favorable winds heated up the ocean surface to record temperatures in August. The average daily water temperature, where the state collects data reached 67.5 F degrees for the month. That was 3 degrees F warmer than any other August since record keeping began in 1905, and 6.5 degrees above average. Farther south on beaches in York and Cumberland counties, lifeguards have measured temperatures exceeding 70 degrees." "I've never seen it as warm as it was and for as long as it was," said Joe Doane, a longtime lifeguard at Scarborough Beach. "It was like bathwater" ("Ocean Temperatures" 2002).

Generally, according to studies conducted by Tim Barnett, a marine physicist at the Scripps Institution of Oceanography in San Diego, as much as 90 percent of greenhouse warming ends up in the oceans. Given the oceans' absorption of heat, said Barnett, whose models have estimated the amount of oceanic warming during the past forty years, "the evidence [of greenhouse warming] really is overwhelming" (Radowitz 2005).

During the spring of 2005, James Hansen and several other scientists published new temperature readings from the deep ocean that trace a clear warming trend indicative of the planet's thermal inertia and energy imbalance—the difference between the amount of heat absorbed by

Earth and the amount radiated out into space. This thermal imbalance (0.85 watts plus or minus 0.15 per square meter) is evidence of a steadily warming world, raising the odds of a catastrophic sudden change marked by rising seas and melting icecaps (Hansen et al. 2005, 1431).

Hansen and colleagues concluded that the unusual magnitude of the warming trend could not be explained by natural variability but instead fit precisely with theories suggesting that human activity is the dominant "forcing agent." "This energy imbalance is the 'smoking gun' that we have been looking for," Hansen said in a prepared summary of the study, which was published in the journal *Science*. "The magnitude of the imbalance agrees with what we calculated using known climate forcing agents, which are dominated by increasing human-made greenhouse gases. There can no longer be substantial doubt that human-made gases are the cause of most observed warming" (Hall 2005, A-1).

By Hansen's estimate, twenty-five to fifty years are required for Earth's surface temperature to reach 60 percent of its equilibrium response—a key concept when diplomacy and policy usually respond to experience rather than to expectations of climate change. Regarding those expectations, 0.85 plus or minus 0.15 watts per square meter of excess heat absorbed by the Earth may not seem like much, but multiplied by many square meters over many years, it adds up.

The Earth's planetary energy imbalance did not exceed more than a few tenths of a watt per square meter before the 1960s, according to Hansen and colleagues. Since then, the excess heat being absorbed by the atmosphere has grown steadily—except for years after large volcanic eruptions—to a level much above historical averages. Much of this excess heat ends up in the oceans, where it helps to melt ice in the Arctic and Antarctic. In the measured tones with which scientists express themselves, accelerating icemelt could "create the possibility of a climate system in which large sea-level change is practically impossible to avoid" (Hansen et al. 2005, 1434). Hansen and colleagues wrote that continuing the present energy imbalance could lead to a climate that is "out of control."

13 SEA-LEVEL RISE

The oceans are the final "stop" in global warming's feedback loop, and potentially one of the most important for human beings—not because we live *in* the oceans, of course, but because more than 100 million people worldwide live within one meter of mean sea level (Meier and Dyurgerov 2002, 350). The situation is particularly acute for island nations. Consider, for example, Indonesia. Jakarta and sixty-nine other sizable cities along Indonesia's coasts will probably be inundated as global warming causes ocean levels to rise during decades to come, according to Indonesia's Secretary of Environment Arief Yuwono ("70 Cities" 2002). Regarding ocean warming, Tom Wigley, a senior scientist at the National Center for Atmospheric Research in Boulder, Colorado, commented: "We're heading into unknown territory, and we're heading there faster than we ever have before" (Revkin 2001, A-15).

By 2005, many oceanographers were hedging their estimates of sea-level rise during the twenty-first century upwards. Richard B. Alley and several other climate-change specialists who specialize in the future of ice and oceans wrote that while sea level may rise about half a meter during the century due to warming climate around the world, "Recent observations of startling changes at the margins of the Greenland and Antarctic ice sheets indicate that dynamical responses to warming may play a much greater role in the future mass balance of the ice sheets than previously considered [and] sea-level projections may need to be revised upward" (Alley et al. 2005, 456).

Ice shelves near the edges of Greenland and Antarctica are shrinking most quickly, even as areas of East Antarctica increase due to heavier snowfalls, also caused at least in part by warming temperatures. Parts of inland Greenland have also been experiencing heavier snowfalls, causing the ice sheet there to grow by 6 to 7 centimeters a year between 1992 and 2003 (Johannessen et al. 2005, 1013). Alley et al. commented that "Ice shelves are susceptible to attack by warming-induced increases of meltwater ponding in crevasses that cause hydrologically driven fracturing and by warmer sub-shelf waters that increase basal melting" (Alley et al. 2005, 458).

HUMAN-PROVOKED WARMING IN THE OCEANS

Marine physicist Tim Barnett, a marine physicist of the Scripps Institution of Oceanography, La Jolla, California, told a meeting of the American Geophysical Union (in San Francisco) during late December 2000 that the upper 9,000 feet of the world ocean shows "a definite sign of human-caused warming" ("Global-Warming Signal" 2000, 4). A team led by Barnett compared measurements of the ocean's heat content with computer models that account for greenhouse gas increases in the atmosphere, as well as other influences on climate. The model accurately mimicked actual observations, which Barnett interpreted as compelling evidence that humans are responsible for at least part of the oceanic warming trend. Barnett said that the findings are "strong enough to overcome his long skepticism about the models' ability to pinpoint a human influence amid all the naturally chaotic ups and downs of climate" (Revkin 2001, A-15). According to Barnett and colleagues, the chance that most of the observed ocean warming is caused by human factors is 95 percent (Barnett, Pierce, and Schnur 2001, 270). "The curves were so good, the data looked faked," said Barnett. "You just don't get it this good the first time around. But there it was" (McFarling 2001, A-1).

Barnett's study was described in the same issue of *Science* (April 13, 2001) as another article detailing results of a study led by Sydney Levitus, director of the National Oceanic and Atmospheric Administration's ocean climate laboratory, in Silver Spring, Maryland. Levitus and colleagues spent ten years collecting and analyzing 2.3 million temperature measurements in the world's oceans to detect anthropomorphic contributions to their temperature levels. The oceans are thought to be a more constant "thermostat" for the Earth than air temperatures, which vary widely day by day. "The oceans are the memory of the climate system," said Levitus (Levitus et al. 2001; McFarling 2001, A-1). These readings from the world's oceans indicate that human influences have been an important cause of a 0.11 degree C warming in the upper two miles of the oceans since 1955. This increase may sound small, but spread across the world's oceans, it contains enough heat energy to meet California's present-day energy requirements for 200,000 years (McFarling 2001, A-1). The two studies also suggest that, even if the emission of greenhouse gases were stopped immediately, the effects of warming already present in the atmosphere would double the rate of warming in the oceans within two to four decades. "We're at the very bottom of a sharply accelerating curve," Barnett said (p. A-1).

Kevin Trenberth, director of the climate analysis section of the National Center for Atmospheric Research, who has written on the use and abuse of climate models, questioned the results of these studies because ocean temperature readings are sparse in some parts of the globe, particularly the southern polar oceans. He added that the two models use different variables. The Parallel Climate model used by Barnett is not fully accurate, Trenberth said, "because it does not include changes in energy from the sun and the impact of volcanoes, both of which have a cooling effect." Levitus' Geophysical Fluid Dynamics Laboratory model may be a bit too warm, Trenberth added (McFarling 2001, A-1).

Some computer models indicate that the average global sea level will rise one to two feet during the twenty-first century because of feedback loops. The oceans are the proverbial "end of the line" of the warming feedback loop. Warming reaches the seas only after a number of decades. Thus, the sea level today may reflect levels of greenhouse gases emitted into the atmosphere forty to fifty years ago. Even if greenhouse gas levels were stabilized today at present levels, sea levels might continue to rise for several hundred years. After 500 years, sea-level rise from thermal expansion may have reached only half of its eventual level, "which models suggest may lie within a range of 0.5 to 2.0 meters to 1.0 to 4.0 meters for carbon-dioxide levels at twice and four times pre-industrial [levels] respectively" (Houghton et al. 2001, 77). Similarly, ice sheets will continue to react over thousands of years, even if greenhouse gas levels stabilize in the atmosphere. An average annual warming of 3 degrees C, sustained over millennia, could melt most of the Greenland ice sheet, eventually raising worldwide sea levels about seven meters.

A taste of what's to come in Greenland was provided late in 2004 through a report in *Nature* that described how Greenland's Jakobshavn Glacier's advance toward the sea has accelerated significantly since 1997. The speed of the glacier nearly doubled between 1997 and 2004 to nearly fifty cubic kilometers a year. This single glacier was responsible for about 4 percent of worldwide sea-level rise during the twentieth century (Joughin, Abdalati, and Fahnestock 2004, 608).

Warming of the oceans has been most notable, relatively speaking, where the sea is usually the coldest, including waters adjacent to Antarctica. A study using seven decades of temperature data indicated that mid-depth water around Antarctica was warming nearly twice as quickly as the oceans' worldwide average (Cowen 2002, 14). Sarah T. Gille's analysis found that the mid-depth water had warmed 0.17 degrees C since 1950 (Gille 2002, 1275). Gille, of the University of California, San Diego, said: "We thought the ocean between 700 and 1,100 meters (2,300 and 3,600 feet) was pretty well insulated from what's happening at the surface. But these results suggest that the mid-depth Southern Ocean is responding and warming more rapidly than global ocean temperatures [generally]." Gille noted that the Southern Ocean "is a very climatically sensitive region" (Cowen 2002, 14). It is at one end of the conveyor-belt circulation that carries heat toward the poles in upper-level currents and returns cold water back toward the equator at great depths—a key part of the system that maintains Earth's present climate. Any change in Antarctic waters could directly affect circulations in the Atlantic, Indian, and Pacific oceans. Ocean circulation already may have been affected. Gille, who reported her findings in *Science*, said that the current has moved closer to the polar continent as the mid-depth water has warmed (p. 14).

Many of the world's shorelines already are eroding, only partially because seas are rising. In many areas, shorelines are subsiding from the last ice age as well as from removal of underground water and oil reserves. The only major exceptions are areas of high sediment supply, such as along the rims of active delta lobes and regions of glacial outwash. Sea level is rising along mid-latitude coastal plain coastlines at typical rates of thirty to forty centimeters per century. Typical rates of shoreline retreat average thirty centimeters to one meter per year (Pilkey and Cooper 2004, 1781).

By one estimate, global sea levels rose by roughly fifteen to twenty centimeters (six to eight inches) during the twentieth century, more from thermal expansion than from melting ice (Miller and Douglas 2004, 406). The rate at which sea level rose during the twentieth century was ten times faster than the average rate for the preceding 3,000 years (Smith 2001, 7).

SEA-LEVEL RISE: "WE'RE LOSING THE BATTLE"

Rising seas, driven by melting ice and thermal expansion, were already swamping beaches, islands, and low-lying coastlines all over the globe as the twenty-first century dawned. At least 70 percent of sandy beaches around the world were receding; in the United States, roughly 86 percent of East Coast barrier beaches (excluding evolving spit areas) have experienced erosion during the last century (Zhang, Douglas, and Leatherman 2004, 41). In some areas, natural subsistence was compounding the problem. Removal of water (and sometimes oil and other resources) from the subsurface also complicated matters. "We're losing the battle," said Stanley Riggs, a geologist at East Carolina University in Greenville, North Carolina (Boyd 2001, A-3). The Outer Banks of North Carolina have been eroding rapidly. "Highway 12 is falling into the ocean. What was once the third row of houses [on the beach] is now the first row," Riggs told a National Academy of Sciences conference on coastal disasters (p. A-3).

A French and American space satellite named *Jason* was launched in 2001 to monitor the upward creep of the seas. In orbit 830 miles above the Earth, *Jason*'s radar altimeter is able to calculate the sea level within an accuracy of one inch, according to Ghasser Asrar, NASA's associate administrator for Earth science (Boyd 2001, A-3). Since 1900, for example, sea levels have risen 12.3 inches in New York City; 8.3 inches in Baltimore; 9.9 inches in Philadelphia; 7.3 inches in Key West, Florida; 22.6 inches in Galveston, Texas; and 6 inches in San Francisco. The rate of sea-level rise has been accelerating over time. At the port of Baltimore, at the head of Chesapeake Bay, for example, the water level crept up at only about one-tenth of an inch per year for much of the twentieth century. After 1989, however, the level rose by half an inch per year, according to Court Stevenson, a researcher at the University of Maryland's Center for Environmental Science (p. A-3). Sea levels have risen twelve to twenty inches on the Maine coast, and as much as two feet along Nova Scotia in 250 years, according to an international team of researchers. Global warming is the main factor, said Roland Gehrels of England's University of Plymouth. He commented that the rate of sea-level rise accelerated during the twentieth century, "as industrialization swept the globe" ("Global Warming Blamed" 2001, 20-A).

Alarm over rising sea levels and subsidence in Shanghai, China's largest city (population 16 million), has prompted officials to consider building a dam across its main river, the Huangpu. "Its main function is to prevent the downtown areas from being inundated with floods," Shen Guoping, an urban planning official, told the *China Daily* ("Shanghai Mulls" 2004). Rising water levels of the Huangpu, provoked by rising sea levels due to global warming as well as subsidence, has resulted in the construction of flood walls hundreds of kilometers in length. At the same time, subsidence caused by the pumping of ground water and rapid construction of skyscrapers has averaged over ten millimeters a year.

A half- to one-meter rise in sea levels could submerge three of India's biggest cities (Bombay, Calcutta, and Madras) by 2020, according to Rajiv Nigam, a scientist with India's Geological Oceanography Division. Nigam said that a one-meter rise in sea level could cause five trillion rupees (U.S. $108 billion) worth of damage to property in India's Goa province alone. "If this is the quantum of damage in a small state like Goa that has only two districts, imagine the extent of property loss in metros like Bombay," Nigam added at a workshop at the National College in Dirudhy, Tamil Nadu state ("Warming Could Submerge" 2003).

By the year 2000, rising sea levels were nibbling up to 150 meters a year from the low-lying, densely populated Nile River Delta. At Rosetta, Egypt, a sea wall two stories high has slowed the march of the sea, which is compounded by land subsidence in the delta; but "sea walls cannot stop the rising [salinity of] the palm groves and fields adjoining the shore" (Bunting 2000, 1). A one-meter rise in sea level could drown most of the Nile Delta, 12 percent of Egypt's arable land and home to 7 million people in 2004.

SEA-LEVEL RISE MAY ACCELERATE

Writing in the March 2004 edition of *Scientific American*, James Hansen, director of the NASA Goddard Institute for Space Studies in New York City, warned that catastrophic sea-level increases could arrive much sooner than anticipated by the Intergovernmental Panel on Climate Change (IPCC) (Holly 2004). The IPCC has estimated sea-level increases of roughly half a meter over the next century if global warming reaches several degrees C above temperatures seen in the late 1800s. Hansen warned that if recent growth rates of carbon dioxide emissions and other greenhouse gases continue during the next fifty years, the resulting temperature increases could provoke large rises in sea levels with potentially catastrophic effects.

Hansen warned that, because so many people live on coastlines within a few meters of sea level, a relatively small rise could endanger trillions of dollars worth of infrastructure. Additional warming already "in the pipeline" could take us halfway to paleoclimatic levels, which raised the oceans five to six meters above present levels during the Eemian period, about 120,000 to 130,000 years ago. Past interglacials have been initiated with enough ice melt to raise sea levels roughly a

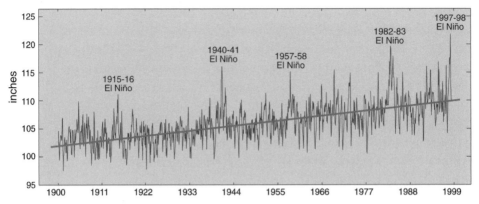

Sea-level measurements collected at Fort Point in San Francisco since before 1900 form the longest continuous sea-level record for any site on the West Coast of North America. Courtesy of the U.S. Geological Survey.

meter every twenty years, "which was maintained for several centuries" (Hansen 2004, 73).

An important issue in sea-level change, wrote Hansen, relates to "the question of how fast ice sheets can disintegrate" (Hansen 2004, 73). "In the real world," wrote Hansen, "ice-sheet disintegration is driven by highly non-linear processes and feedbacks." For example, higher sea levels can physically lift marine ice shelves that formerly prevented land ice sheets from sliding into the ocean. This effect accelerates the breakup of the land ice (Holly 2004). In addition, melting glacial waters flow downward through holes in the ice to the bottom of the ice mass, where they become a lubricant that further accelerates the disintegration of the land ice and its flow into the sea.

Although buildup of glaciers is gradual, "once an ice sheet begins to collapse, its demise can be spectacularly rapid" (Hansen 2004, 74). The darkening of ice by black-carbon aerosols (soot)—pollution associated with the burning of fossil fuels—also accelerates melting. Although the timing of melting is uncertain, wrote Hansen, "global warming beyond some limit will make a large sea-level change inevitable for future generations" (p. 75). Hansen estimated that such a limit could be crossed with about 1 degree C of additional worldwide warming. This amount is below even the most conservative estimates of the IPCC for the next fifty years. Hansen recommended the restriction of methane and soot emissions to balance slow growth in carbon dioxide. That, plus improved energy efficiency and increased use of renewable energy sources, may buy time. He added that new technologies might be developed "that we have not imagined." The question, he concludes, is, "Will we act soon enough?" (p. 77).

Global average surface temperature has risen about 0.75 degree C since a worldwide temperature network became established in the late 1800s, with most of the increase, about 0.5 degree C, after 1950. About 70 percent of the increase in anthropogenic greenhouse gases occurred after 1950. The Earth already has absorbed 0.4 to 0.7 degree C worth of warming that is not yet reflected in the atmosphere because of delayed feedback, mainly a lag in ocean warming (Hansen 2004, 72). Hansen warned that while generally moderate warming thus far "leave[s] the impression that we are not close to dangerous anthropogenic interference, I will argue that we are much closer than is generally realized, and thus the emphasis should be on mitigating the changes rather than just adapting to them" (p. 73).

WARMING, RISING SEA LEVELS, AND THE SWAMPING OF COASTAL AREAS

A report by the World Resources Institute (WRI) has warned that, if the Earth's coastal zones continue to flood, their ability to provide fish, protect homes and businesses, reduce pollution and erosion, and sustain biological diversity will be gravely endangered. "Unless things change very quickly, the world's coastal areas face a grim future. Many important coastal habitats like lagoons, wetlands, mangroves, and coral reefs are disappearing," said Jonathan Lash, WRI president, during the release of the report "Pilot Analysis of Global Ecosystems (PAGE): Coastal Ecosystems" (Amor 2001).

Nearly 30 percent of the world's coastal ecosystems have already been extensively altered by growing demands for human housing, industry, and recreation. An estimated four of every ten people worldwide now live within 100 kilometers of a coast. "Coastal populations are exploding, and as they increase, pressures on coastal ecosystems will follow," said Lash (Amor 2001). Nearly two-thirds of all the fish harvested in the world depend on coastal wetlands, mangroves, sea-grass beds, and coral reefs; about 95 percent of the world's marine fish harvest comes from coastal waters. More low-value fish are being caught today as stocks of valuable fish such as cod, hake, and tuna decline; roughly 75 percent of fish stocks are being depleted or fished at their biological limits. Sea levels that may rise by as much as ninety-five centimeters at the end of the twenty-first century could produce storm surges that will intensify erosion, habitat loss, increased salinity of fresh-water aquifers, and extreme coastal flooding, said Yumiko Kura, one of the lead authors of the report (2001).

PROSPECTIVE CLIMATE CHANGES ALONG THE GULF OF MEXICO COAST

Global warming will have serious long-term effects on the U.S. Gulf Coast, including more frequent flooding, worse droughts, and a diminishing supply of fresh water, according to a study by a team of ten scientists with expertise in the region. The study detailed potential consequences during the next 50–100 years of higher temperatures and rising sea levels on the Gulf coastal plain, from the Laguna Madre to the Florida Keys (Freemantle 2001, 32).

The Gulf Coast region is disproportionately susceptible to changes in temperature and sea level, because a small increase in temperature in this humid area can result in a drastic rise in the area's July heat index. A small rise in sea level poses serious problems in a coastal region that is flat, overdeveloped, and slowly sinking even without sea-level rise and thermal expansion provoked by global warming. This study said that sea level may rise between eight and twenty inches during the twenty-first century. Adding sea-level rise to ground subsidence, coastal residents can expect a net sea-level rise of fifteen to forty-four inches. At the same time, many scientists' forecasting models agree that precipitation will become more uneven. Extreme rainfall events, such as the one that caused flooding in the Houston area in June 2001 from the remnants of tropical storm Allison, are expected to occur more frequently (Free-mantle 2001, 32).

Joseph Siry, director of the Florida Climate Alliance, called the report "an early-warning advisory about a problem that shows no signs of going away" (Hollingsworth 2001, 1). "The potential impacts are so great [that] we must plan now to avert them," he said. The studies suggest that accelerated warming of the global climate might lead to a catastrophic rise in local sea level, dramatic weather fluctuations, and an increase in heat-related and insect-borne illness (p. 1).

Florida has already been experiencing early effects of global warming in the form of retreating shorelines, dying coral reefs, more wildfires, and salt water seeping into the freshwater aquifers, according to a peer-reviewed study by scientists from Florida universities (Hollingsworth 2001, 1). Such studies are supported by anecdotal observation. My family, for example, has been living at or visiting the Daytona Beach area for half a century. In my childhood, during the 1950s, the beach was broad and firm, even at high tide; stock-car racing in the United States was born on that beach. By the year 2004, sea-level rise and the general sinking of the land from withdrawal of water, along with the erosive effects of three hurricanes in six weeks, had stripped away much of the beach. The dunes had lost several feet from storm surges, and, at high tide, waves lapped at many bulkheads.

Maximum summer temperatures in Florida could increase 3 to 7 degrees F, boosting July's heat index (the combined effect of heat and humidity) by 10–25 degrees. "We're already doomed to having the consequences of global climate change no matter what we do," said Mark Harwell, a professor of marine ecology at the University of Miami. "But we can significantly reduce the consequences if we acknowledge and plan for them now" (Hollingsworth 2001, 1). About 1,000 people move into Florida every day, a pace expected to more than double the population to 34 million in thirty years. "Water supply is critical to this scenario. It's going to be the problem for Florida in the future," said Susanne Moser of the Union of Concerned Scientists (p. 1).

An anticipated sea-level rise of eight to twenty inches during the twenty-first century could imperil houses that were within 200 feet of the shoreline in 2000, which may be below the high-tide line by the year 2100. A panel of scientists whose members spent two years studying the impact of global warming on Florida and other Gulf Coast states announced that rising sea levels are likely to inundate many coastal areas, including Tampa Bay, during coming decades. Rising seas will force Floridians to build extensive sea walls to protect their waterfront homes, condominiums, and businesses, the scientists said. The disappearance of the beaches is likely to cost Florida's tourism industry dearly. The state's major real-estate assets are its beaches, now framed by ranks of condominiums. Ricardo Alvarez of the International Hurricane Center at Florida International University said during a news conference announcing the study results, "we sell good beaches, climate and sunshine" (Pittman 2001, 3-B).

Higher storm surges also will probably worsen the impact of future hurricanes in Florida and other states bordering the Gulf of Mexico. The Gulf Coast climate study warned that, because of rising seas, saltwater intrusion is likely to contaminate more drinking supplies in Texas, Louisiana, Mississippi, Alabama, and Florida. The report, "Confronting Climate Change in the Gulf Coast Region," was sponsored by the Union of Concerned Scientists and the Ecological Society of America. A more concise report, "Feeling the Heat in Florida," was sponsored by the Natural Resources Defense Council.

Parts of coastal Louisiana and Mississippi could lose as much as one foot of elevation within ten years according to a National Oceanic and Atmospheric Administration (NOAA) National Geodetic Survey. The NOAA researchers have warned that populated areas will face increased dangers from storm surges and flooding due to ongoing subsidence of coastal areas along the northern Gulf of Mexico. Coastal wetlands in Louisiana have been disappearing at the rate of thirty-three football fields per *day* (Bourne 2004, 89). The NOAA researchers estimate that, at the present rate of subsidence, 15,000 square miles of land along the southern Louisiana coast will subside to sea level or below within the next seventy years ("Coastal Gulf" 2003). Shoreline in this area is sinking due to natural processes as well as the withdrawal of subsurface oil and water. In southern Louisiana, roughly 1 million acres of coastal marsh were converted to open water between 1940 and 2000, with losses accelerating over time with the quickening pace of sea-level rise (Inkley et al. 2004, 8, 13). Subsidence adds to the effects of sea-level rise due to rising temperatures. The state of Louisiana lost about 1,900 square miles of coastal marshes, an area the size of Delaware, during the twentieth century. It may lose another 700 square miles by 2050 at present rates of erosion and subsidence. By that time, a total of one-third of the state's coast will have washed into the Gulf of Mexico since the early twentieth century.

Matt Crenson (2002) of the Associated Press sketched a picture of coastal erosion's practical effects:

> A few decades ago Isle de Jean Charles was a patch of high ground in a sea of grassy marsh teeming with catfish and crawdads. Today the small community is a true island, regularly flooded during storms and sometimes even at high tide. In a few years it will be submerged completely. Deme Naquin, 75 years old, remembers paddling a flat-bottomed pirogue to school as a boy. Now he's getting ready to leave the only place he has ever called home. The U.S. government has offered to resettle the island's 270 residents because a new hurricane protection plan leaves them outside the ramparts.

"Another hurricane and the road's going to be gone," said Chad Naquin, Deme's twenty-nine-year-old grandson. "It would be hard to leave, but in the long run it would be the best thing" (Crenson 2002). As many as thirty-five square miles of Louisiana's wetlands sink into the Gulf of Mexico each year. In some places, the coastline has retreated as much as thirty miles.

Rising sea levels provoked by global warming, combined with Louisiana's subsiding coastline, could raise the Gulf of Mexico adjacent to Louisiana by more than three feet in the next century, drowning remaining wetlands and increasing the risk of hurricane storm surges for New Orleans and coastal towns, according to some scientific reports. Increasing temperatures will also result in dangerously hot summers, warmer winters, and less rainfall along the Gulf Coast. Expected warming may raise summertime high temperatures in Louisiana by 3 to 7 degrees F and wintertime low temperatures by 5 to 10 degrees F (Schleifstein 2001, 1).

"Here, in Louisiana, we're particularly vulnerable to climate change because we're already living on the edge," said Denise Reed, a wetlands scientist at the University of New Orleans and one of the report's authors. "Many of our coastal communities, Dulac, Irish Bayou, Yscloskey, already have water lapping at their roads. Add a foot of water from climate change, and you're facing more than an inconvenience" (Schleifstein 2001, 1).

Most states along the Atlantic and Gulf coasts of the United States have been experiencing problems similar to Louisiana's. "We're not going to be the only ones in the boat," said Al Naomi, a project manager in the New Orleans District of the U.S. Army Corps of Engineers. "We're just in the boat first" (Crenson 2002). Loss of coastal wetlands may devastate Louisiana's $2 billion-a-year fishing industry. Many fish, crabs, shrimp, and other marine animals rely on wetlands as a nursery, where their young feed. By some calculations, Louisiana's wetlands produce as much as 40 percent of the seafood caught in the United States (Crenson 2002).

Sinking land and rising waters imperil coastal residents as well as wildlife. Before Hurricane Katrina hit the area in 2005, the Louisiana Red Cross had estimated that between 25,000 and 100,000 people could become refugees if a major hurricane hit New Orleans from the southeast, driving the waters of Lake Pontchartrain into the city. Many of their homes already lie below mean sea level, often behind dikes. The estimated scope of the disaster in this case was greatly underestimated. According to Crenson, the Red Cross is so pessimistic about southern Louisiana's prospects in the face of a major storm that it closed all of its relief centers south of Interstate 10, which runs across the state from Lake Charles to New Orleans. The Red Cross said it would be foolhardy to operate disaster centers in an area that people should abandon in a hurricane (Crenson 2002). Ivor van Heerden, deputy director of the Louisiana State University Hurricane Center, described a horrific ordeal for anybody riding out a major hurricane in New Orleans. "If you survive the flying debris and your house collapsing," van Heerden said, "then you're going to have to deal with a minimum of 13 feet of water" (2002).

The weeks or months after a major hurricane could present New Orleans residents with other problems. The city sits in a bowl ringed by protective levees, so water would sit for weeks or months until officials could find a way to breach the barriers to drain it. Stagnant pools could be contaminated with toxic waste from dozens of petrochemical plants that line the Mississippi, as well as human waste and decomposing carcasses. "This is a $50 billion disaster," van Heerden said, surveying expensive new housing developments on the shores of Lake Pontchartrain (Crenson 2002).

During the end of August 2005, Hurricane Katrina tested the emergency-preparation modeling for New Orleans and the rest of the Central Gulf coast with 150-mile-an-hour winds, a storm surge as high as forty

Floodwaters from Hurricane Katrina flowed over a levee along the Inner Harbor Navigational Canal near downtown New Orleans, August 30, 2005. © AP/Wide World Photos.

feet, and the second-lowest barometric-pressure reading in U.S. history. A majority of the city's half-million residents evacuated on August 27 and 28, as forecasters raised a possibility that the huge storm surge would swamp the city's levy system, mixing floodwater with sewage, turning large parts of the city into a huge, festering open-air toilet. Tens of thousands who could not leave the city gathered in the Superdome. The storm was stoked by a Gulf of Mexico at nearly 90 degrees F.

By August 30, two days after Katrina came ashore over Gulfport, Mississippi, the Federal Emergency Management Agency (FEMA) was calling its landfall the most significant natural disaster in the history of the United States. Eighty percent of New Orleans was under water, and the city had no power, no drinking water, and no place to bury the uncounted dead. Rescuers passed the dead to rescue the living, and authorities ordered those who had ridden out the storm (and survived) to leave. Along the Gulf Coast, in and near Biloxi, Gulfport, and Mobile, a thirty- to forty-foot storm surge had turned entire beachfront towns into piles of broken bricks and kindling. The fetid, humid heat was turning what remained into a stinking health hazard. A storm surge estimated at forty feet had wiped away the town of Waveland, Mississippi, fifty miles northeast of New Orleans. Large parts of Biloxi, Gulfport, and other coastal cities and towns suffered damage on an apocalyptic scale. Five million people lost power, many of them for many weeks. The "worst-case scenarios" paled beside reality.

Three days after the hurricane hit, chaos gripped New Orleans as looters ran wild, food and water supplies dwindled, and bodies floated in the floodwaters. Evacuation of the Superdome began, as the leaking shelter became a festering hellhole without water, power, food, or toilets. Officials said there was no choice but to abandon the city, perhaps for several months. The landfall of Katrina provoked an unprecedented long-term exodus, the largest in U.S. history, as 1 to 2 million environmental refugees sought refuge outside the affected area. Researchers who flew over 180 miles of coast between Pensacola, Florida, and Grand Isle, Louisiana, said that along the shore, for blocks inland, nothing remained but concrete slabs and chunks of asphalt. Often, they said, it was impossible to tell what had been there before the storm.

On the fourth day after landfall, in New Orleans, fights and fires broke out, corpses lay out in the open, and rescue helicopters and law enforcement officers were shot at by snipers as New Orleans descended into anarchy. City officials issued "a desperate SOS," as the city was evacuated. Anger mounted across the ruined city, with tens of thousands of storm victims increasingly hungry, desperate, and tired of waiting for buses to take them out. Bodies lay by roadsides, eaten by rats.

I woke Friday morning, September 2, and turned on the Cable News Network, to news of chemical explosions and fires in New Orleans. People were being raped and shot on the streets, as hundreds of bodies festered in the floodwaters. Many police had turned in their badges and walked away. A police officer of eighteen years was being interviewed on CNN as I dressed. He was asked to characterize his city. "It's gone," he said, breaking down in tears.

Tim Wagner, Nebraska State Insurance Commissioner, was interviewed in the Omaha *World-Herald* three days after Katrina hit, saying that global warming is causing weather-related disasters to be more severe and more frequent. "It's scary," he said. "It's becoming apocalyptic" (Jordan 2005, 7-A).

Saturday, September 3, the sixth day after the storm's landfall, the Associated Press reported: "By mid-afternoon, only pockets of stragglers remained in the streets around the convention center, and New Orleans paramedics began carting away the dead. A once-vibrant city of 480,000 people, overtaken just days ago by floods, looting, rape and arson, was now an empty, sodden tomb" (Breed 2005). Shopping centers and rows of warehouses burned out of control, filling the sky with acrid black smoke. Huge cruise ships and domed stadiums in neighboring states (such as the Houston Astrodome) were turned into homeless shelters.

A week after the storm hit, rescue crews began to haul in bodies. The *New York Times* reported: "Seven days after Hurricane Katrina devastated the Gulf Coast, the New Orleans known as America's vibrant capital of jazz and gala Mardi Gras celebrations was gone. In its place was a partly submerged city of abandoned homes and ruined businesses, of bodies in attics or floating in deserted streets, of misery that had driven most of its nearly 500,000 residents into a diaspora of biblical proportions. . . . Officials warned of an impossible future in a destroyed city without food, water, power, or other necessities, only the specter of cholera, typhoid, or mosquitoes carrying malaria or the West Nile virus" (McFadden 2005).

By September 10 New Orleans police, warning of disease risks from increasingly fetid waters, were removing about 10,000 remaining residents. Never before had a large city in the United States been emptied of its people. Two weeks passed between the storm and a concerted effort to collect bodies languishing in the gradually declining floodwaters. Years before, scientists in the area had modeled the same situation in an exercise they called "Hurricane Pam." They had known that New Orleans would flood with massive loss of life, but the gruesome nature of the reality had escaped even them.

Two and a half weeks after Katrina struck, President George W. Bush stood in the French Quarter and told a national television audience that his government, which had been severely criticized for its tardy response to the storm, would "do what it takes" to rebuild New Orleans and the rest of the devastated Gulf Coast. Estimates of the cost at the time ranged up to $200 billion. The next day, in *Science*, P. J. Webster and colleagues linked rising water temperatures directly to the number, duration, and intensity of hurricanes (Webster et al. 2005, 1844–1846). Thus, with its land base subsiding and seas continuing to slowly rise, New Orleans faces the future in what amounts to a game of climatic Russian Roulette, with the size of the gun and the number of bullets slowly increasing. Bush said nothing about that.

The number of storms in the two most powerful categories, 4 and 5, rose to an average of eighteen a year worldwide since 1990, up from eleven in the 1970s, according to Webster's report. The most powerful tropical cyclones worldwide increased in number 80 percent between 1970 and 2005, a time of steadily rising temperatures (Kerr 2005, 1807). Rising numbers of people living in coastal areas has also been increasing the amount of damage from the intense storms that strike land. Warming the oceans is like "throwing more wood on the fire" (1807). However, wrote Webster and colleagues, wide variations in the number of cyclones year to year make any straightforward association between sea-surface temperatures and their number, duration, and intensity statistically difficult (Webster et al. 2005, 1844). Wind sheer patterns and levels of moisture, among other factors, also play important roles in the strength of tropical cyclones.

Less than four weeks after Katrina, another powerful hurricane, Rita, formed over the southern Bahamas, brushed southern Florida, and then rapidly intensified to a Category 5 with top sustained winds of 175 miles an hour over the 85 degree F waters of the Gulf of Mexico, before it slammed into the Texas-Louisiana border as a Category 3. In addition to the other factors imperiling the hurricane-prone Gulf Coast, the loop current, a great ribbon of very warm water meandering through the Gulf of Mexico, is another key factor energizing hurricanes. This current provided heat energy that helped transform Hurricanes Rita and Katrina into a rare single-season pair of Category 5 storms. The current carries warm water from the Caribbean Sea around the western horn of Cuba into the gulf, where it helps spawn the Gulf Stream.

Hurricane Katrina became the costliest hurricane in U.S. history despite the fact that it weakened to a Category 3 storm, with winds of 125 miles an hour, shortly before landfall. Thus, most of Katrina's damage was wrought by storm surge and flooding rather than wind. Winds in New Orleans, to the west of the storm's center, were probably even weaker than that, at Category 1 or 2 speeds, according to a report issued by the National Hurricane Center in late-December 2005. "This is a further indictment of the levee system," Ivor Van Heerden, a Louisiana State University professor and leader of a team of Louisiana investigators probing the cause of the levee breaches. "It indicates that most of the flooding of downtown New Orleans was a consequence of man's folly" (Whoriskey and Warrick 2005, A-3). "The water level in the canals wasn't that high when the floodwalls breached," according to Robert Bea, a civil engineering professor at the University of California at Berkeley and a member of an investigating team funded by the National Science Foundation. "We had a premature failure of the defense system" (p. A-3).

Hurricanes Katrina and Rita alone submerged 100 or more square miles of formerly dry land in Louisiana. Since the 1930s, Louisiana has lost almost 2,000 square miles (an area about the size of Delaware) to encroaching seas. Taken as an average, that would be between six and seven linear miles along a 300-mile coastline.

The Louisiana coast's problems do not end with problems provoked by withdrawal of oil and water, oil-industry canals, slowly rising seas, and more intense storms that arrive more frequently. In addition, the construction of levees and other flood-control measures deprive the area of silt, nature's way of nourishing wetlands. "The whole surface is sinking," said Abby Sallenger, a scientist with the U.S. Geological Survey. "It's almost changing before your eyes. It's grassland turning to open water [as] the ponds turn into lakes" (Dean 2005).

The infamous hurricane season of 2005, which had begun so early, also ended late, as Hurricane Wilma blew up from a tropical storm to a Category 5 overnight October 18 and 19 over warm open water east of Honduras, setting a record for lowest barometric pressure (884 millibars) in the Atlantic basin—lower than Katrina's 902. Maximum sustained winds exploded from 80 to 175 miles an hour in one night. Wilma also set another record, as the first hurricane in recorded history to erupt from a Category 1 to a Category 5 in less than twenty-four hours. Three of the ten most intense hurricanes in the history of the Atlantic basin had occurred in one year.

Students of atmospheric science have been taught that hurricanes take about five days to mature. Not this one. As Wilma devastated Northeastern Yucatan (including Cancun and Cozumel), a new storm, Alpha, formed south of Hispaniola, breaking the record for number of hurricanes in one season. The record in the Atlantic basin reaches back about 150 years. Alpha later was absorbed by the stronger wind-field of Wilma after that storm had ripped across southern Florida as a Category 3, with strongest sustained winds of 125 miles an hour. In Cancun, most of the resort's white-sand beaches were stripped down to limestone rocks. Even later the same month, on October 30 and 31, Hurricane Beta pillaged Nicaragua with 110-mile-an-hour winds and torrential rains.

CHESAPEAKE BAY'S MARSHES DISAPPEARING

Richard A. White once had planned to live the rest of his life in his waterfront home on Chesapeake Bay, which, according to Tom Pelton's account in the *Baltimore Sun*, "perches at the tip of a filament of land reaching out into Chesapeake Bay and boasts a 50-foot-long veranda with panoramic views of the sunset" (Pelton 2004, 1-A). The sixty-year-old historian has been forced to drink bottled water because salt water has ruined his well. Whitecaps often froth across his lawn, which has shrunk by about forty feet in the past eighteen years. During high tides in the fall, wrote Pelton, White "pulls on tall boots, parks his car a block away on high ground and ties a rowboat beside his door, as his century-old house becomes a tiny island unto itself" (p. 1-A).

White is not alone. More than half of Chesapeake Bay's marshes are in danger of becoming open water, according to a University of Maryland study. Rising sea levels are flooding the marshes that filter pollutants in ground water, guard against erosion, and provide habitat for ducks, geese, and other animals. Michael Kearney, an associate professor at the University of Maryland, College Park, said the marshes could be gone in twenty to thirty years. He said the time line could be shortened if nor'easters become more severe. Kearney and colleague J. Court Stevenson (also of the University of Maryland) said that water levels in the bay generally have been rising since the year 1,000 CE, but that the rate of rise accelerated sharply (to more than ten times the previous rate) during the 1990s. Sea-level rise in the area is being made worse by subsidence of the land (McCord 2000, 1-B).

At the Blackwater National Wildlife Refuge near the shores of Chesapeake Bay, rising waters are rapidly destroying marsh habitat. "Blackwater gives us an example of what will probably occur in a lot of low-lying areas as global warming proceeds and water levels continue to rise," said Stevenson, who has been studying the swampy reserve for more than twenty years. "If we don't do anything about global greenhouse emissions, up to a third of this county where we're standing now will eventually become open water" (Reuters 2001, 11).

Blackwater includes about 6,900 hectares of wetlands, woodlands, and croplands. Of that, 2,800 hectares of marshland is already flooded, and the rising waters claim another fifty hectares each year. According to a Reuters report, "The rivers which run through the refuge are turning salty. Tiny island sanctuaries for bald eagles are disappearing and pines and grasses are dying. The process is accelerated by a gradual sinking of the land mass, over-pumping of underground water and over-grazing" (2001, 11).

Conditions similar to those at Blackwater may soon impact much of the U.S. eastern coastline. The U.S. Environmental Protection Agency (EPA) has said that the sea level is rising more quickly along the eastern U.S. coastline than the worldwide average, and most quickly of all in Chesapeake Bay. Along the Atlantic Coast and the Gulf of Mexico, the EPA forecasts a thirty-centimeter rise in sea level by 2050. That estimate, however, is optimistic; the EPA admits that waters could rise that much by 2025. Given such a sea-level rise, large parts of low-lying cities such as Boston, New York, Charleston, Miami, and New Orleans would become vulnerable to regular flooding (Reuters 2001, 11).

RISING WATERS NIBBLE AT NEW JERSEY AND LONG ISLAND

Rising waters are nibbling at the coastlines of New Jersey and Long Island; some beachfront vacation homes on Long Island have been raised on stilts as tides have risen. Hoboken subway stations have been reinforced with "tide guards" (Nussbaum 2002, A-1). Vivian Gornitz, a sea-rise specialist at NASA's Goddard Institute of Space Studies in New York City, said that the amount of land lost to the sea could accelerate to between two and five times the present rate by the end of the twenty-first century. She said that seas in the area could rise between 4 and 12 inches by 2020, 7 inches to 2 feet by 2050, and 9.5 inches to almost 4 feet by 2080.

Jim Titus, project manager for sea-level rise at the U.S. Environmental Protection Agency, said that such a rise could imperil all beaches in the area. "The data doesn't lie," he said. "The sea is rising. The shore is eroding because the sea is rising. There's no doubt about that." At Long Beach Island, said Titus, sea levels already have risen a foot in seventy years. Two days a month, high tides at full moon (called "spring tides") often cause flooding in the area. Homes are now built on stilts, and garbage cans are anchored to prevent them from floating away on the tides (Nussbaum 2002, A-1). Local officials say that future storms could flood transportation infrastructure, including the Newark-Liberty International Airport, which lies less than ten feet above present sea levels.

"MANAGED REALIGNMENT" ALONG ENGLAND'S COASTLINE

Parts of England's coastline are afflicted by the same problems as the U.S. East and Gulf of Mexico coasts. The land is subsiding, as icemelt and thermal expansion slowly raise sea levels. The U.K. Climate Impact Programme, a government-funded program at Oxford University, forecasts that the sea could be as much as a meter higher by late in the twenty-first century. In addition to climate change, "isostatic rebound," the rise of Scotland's coast following the last ice age, is contributing to a subsiding coastline southward along the English coast. As the sea rises three millimeters a year abreast of Essex, the land itself is sinking half as rapidly. By 2003, the rising waters were threatening the closed Bradwell Nuclear Power Station ("Rising Tide" 2003, 4).

Anthony Browne commented in the *London Times*: "A thousand years after King Canute showed that man could not hold back the tide, the Government has come to the same conclusion. The Environment Agency, the government body responsible for flood defenses, is planning a strategic withdrawal from large parts of the English coastline because it believes that it can no longer defend them from rising sea levels, the result of global warming" (2002, 8).

The new strategy, officially called "managed realignment," will allow the sea to flood low-lying farmland rather than fending off the invading waters by building ever-higher defenses. The policy, which will allow the encroaching sea to submerge several thousand acres of land, has been welcomed by environmentalists. Farmers, however, argue that the strategy is "unviable" and have demanded more concrete flood defenses. The area of affected coast ranges from the Humber Estuary, around East Anglia, to the Thames estuary and west to the Solent. Strategic withdrawal also has been planned for sections of the Severn Estuary. The first site surrendered to the sea was in Lincolnshire. About 200 acres of farmland were flooded by seawater at Freiston Shore after the flood defense banks were breached to create a salt-marsh bird reserve (Browne 2002, 8).

"Managed realignment is in its infancy . . . but it has been an emotive issue; a lot of people are concerned about it. But you have to get the message across that we are defending property," said Brian Empson, flood defense policy manager for the Environment Agency (Browne 2002, 8). The land that was returned to the sea in 2002 had been reclaimed 150 years earlier and protected by man-made banks. A grass-covered wall in Abbots Hall farm country on the east coast of Essex, which had held back the sea for almost 400 years, was also breached intentionally in 2002 to surrender the area to the sea. The use of breakwaters made of clay, bricks, and, finally, large blocks of concrete, had not stopped the rising waters.

Until recently, the Thames River barrier, built to protect London and surrounding areas from unusually high tides and storm surges, was used an average of two or three times a year. Between November 2001 and March 2002, however, the barrier was raised twenty-three times. A British report released in September 2002 said that 59,000 square miles (home to 750,000 people) in and around London are vulnerable to flooding because they are below high-tide levels, some by as much as twelve feet.

SEA-LEVEL RISE AND SCOTLAND'S ST. ANDREWS GOLF COURSE

Golf has been played at St. Andrews since the fifteenth century, but by the year 2000, global warming, rising tides, and coastal erosion were allowing the sea to swallow some of the game's oldest links. According to one observer, "One theory even claims that golfers will be lucky to enjoy another 50 years of competition on the links in Fife before the Old Course becomes just another submerged attraction akin to the lost city of Atlantis" (Aitken 2001, 20). Peter Mason, external relations manager of the St. Andrews Links Trust, explained how a combination of high tides, large waves, and strong winds has battered the land along the Eden estuary and swept away a coastal path. "Three meters of linksland, just centimeters away from the par-4 eighth hole on the Jubilee Course, were lost to the sea in a single day last year after storms. It fell away so fast, it frightened the life out of us," Mason recalled (p. 20).

Parts of the St. Andrews links once were reclaimed from the ocean. The original courses were called "links" because they were located on strings of dunes and knolls that linked sea and land (Hale 2001, 3-C). Today, all golf courses are called "links" after St. Andrews' origins, whether or not they border water. Satellite surveys indicate that the height of the waves striking Britain's coasts has increased 25 percent in twenty years; hence, some of these holes may be relocated for the first time in their two-century history.

Sloping gabions (stone-filled wire cages) have been installed along a 100-meter stretch where the path ran prior to the storms. This defense was backed by 4,400 cubic meters of sand; an additional 12,000 cubic meters of sand was added later. If that doesn't hold back the sea, more gabions may be needed (Aitken 2001, 20). Roughly 200,000 rounds of golf are played at St. Andrews annually, a major tourist draw. Golfers who drive off the eighth tee of the Jubilee course now stare down at 400 yards of rock-filled cages that stand between the course and the ocean. The third hole on the Royal Aberdeen course once was sixty yards from dunes that marked the ocean's edge. By the year 2001, on the eve of the 130th British Open on the course, golfers teed off at this hole two yards from a forty-foot drop into the ocean (Hale 2001, 3-C). Violent winter storms are responsible for much of the erosion.

Other British golf courses also are eroding. At the Royal West Norfolk Golf Club, near Brancaster in England, advancing seas have isolated the clubhouse, forcing the use of knee-high rubber boots.

RISING SEAS AND THE NETHERLANDS

The Dutch fear that rising storm surges could inundate much of the Netherlands, large areas of which have been reclaimed from the sea. Fears have been expressed that the country's western provinces may flood. The Hague may become uninhabitable as low-lying suburbs of Amsterdam return to marshland or open water.

The Dutch have already been forced to plan the surrender of 200,000 hectares of farmland to river floodplains. A major construction program of floating homes has started. Pieter van Geel, the Dutch environment secretary, said, "Half of our country is below sea level and so beyond a certain level, it is not possible to build dikes anymore. If we have a sea level rise of two meters, we have no control, no possibility of solving that. It's unthinkable. I fear the problem is going to catch us within 25 years, and that is a very short period" (Evans-Pritchard 2004, A-6).

SEA-LEVEL RISE AND PACIFIC ISLANDS

Patrick Barkham of the *London Guardian* reported from the South Pacific that islands widely imagined as paradises are falling victim to rising seas:

> The dazzling white sand and dark green coconut palms of Tepuka Savilivili were much like those on dozens of other small islets within sight of Funafuti, the atoll capital of Tuvalu. But shortly after cyclones Gavin, Hina and Kelly had paid the tiny Pacific nation a visit, islanders looked across Funafuti's coral lagoon and noticed a gap on the horizon. Tepuka Savilivili had vanished. Fifty hectares of Tuvalu disappeared into the sea during the 1997 storms. The tiny country's precious 10 square miles of land were starting to disappear. (2002, 24)

Nothing living remains on the few square meters of sand that used to be the island, "only several odd flip-flops and a rusty tin." Vasuaafua, another small island, had been reduced by the rising ocean to nine coconut palms on a narrow ribbon of sand by 2002. Several years earlier, the island had a sandy beach several hundred yards long. During the highest tides of 2001, the island's meteorological office, near the airstrip on Funafuti, was swamped by seawater (Barkham 2002, 24). Before the mid-1980s, flooding tides usually arrived only in February. By 2002, however, floods became likely at any time between November and March.

Many of these islands face two compounding problems: the sea is rising, while the land itself is slowly sinking as sixty-five-million-year-old coral atolls reach the end of their life spans. The atolls were formed as formerly volcanic peaks sank below the ocean's surface, leaving rings of coral. For five years, the government of Tuvalu has noticed many such troubling changes on its nine inhabited islands. It has thus concluded that, as one of the smallest and lowest-lying countries in the world, it is destined to become among the first nations to be sunk by a combination of global warming—provoked sea-level rise and slow erosion. The evidence, including forecasts by international scientists of a rise in sea level of as much as eighty-eight centimeters during the twenty-first century, has, according to Barkham's report, convinced most of Tuvalu's 10,500 inhabitants that rising seas and more frequent violent storms are certain to make life unlivable on the islands, if not for them, then for their children (2002, 24).

Residents of the islands have been seeking higher ground, often in other countries. The number of Tuvalu's residents living in New Zealand, for example, doubled from about 900 in 1996 to 2,000 in 2001, many of them fleeing the rising seas on their home islands. A sizable Tuvaluan community has grown up in West Auckland (Gregory 2003). The highest point on Tuvalu is only about three meters above sea level. "From the air," wrote Barkham, "its islands are thin slashes of green against the aquamarine water. From a few miles out at sea, the nation's numerous tiny uninhabited islets look smaller than a container ship and soon slip below the horizon" (2002, 24).

"As the vast expanse of the Pacific Ocean creeps up on to Tuvalu's doorstep, the evacuation and shutting down of a nation has begun," Barkham wrote. "With the curtains closed against the tropical glare, the

At top, a Tuvaluan house perches over an empty "borrow pit" dug by United States forces during World War II in order to build the airstrip on Funafuti Atoll, February 22, 2004. At bottom, the same house flooded at high tide. © Torsten Blackwood/AFP/Getty Images.

prime minister, Koloa Talake, works in a flimsy Portakabin at the lagoon's edge on Funafuti." Talake, "who sits at his desk wearing flip-flops and bears a passing resemblance to Nelson Mandela," likens his task to the captain of a ship: "The skipper of the boat is always the last man to leave a sinking ship or goes down with the ship. If that happens to Tuvalu, the prime minister will be the last person to leave the island" (2002, 24).

Many Pacific island farmers report that their crops of swamp taro (*pulaka*), a staple food, are dying because of rising soil salinity (Barkham 2002, 24). Another staple food, breadfruit (*artocarpus altilis*), is also threatened by saltwater inundation. Breadfruit are harvested from large evergreen trees, with smooth bark and large, thick leaves, that reach a height of twenty meters (about sixty feet). The fruit, which is large and starchy, reaches edible maturity once or twice a year, depending on the type of tree. Some varieties bear fruit all year. Breadfruit probably originated in the Moluccas, the Philippines, and New Guinea. Diana Ragone, director of science at the National Tropical Botanical Garden in Hawaii, said that "on some atolls now the people are seeing total die-back of the breadfruit trees, one hundred percent" (Field 2002). Shallow-rooted breadfruit trees are especially vulnerable to increasing numbers of tropical storms, Ragone said. Two cyclones, Val and Ofa, totally wiped out breadfruit trees in parts of Samoa during 2002.

Living within a few meters of ocean level, island residents share a special fear of typhoons that roil the seas and drive storm surges through their villages. Two decades ago, one or two serious tropical storms usually hit the islands each decade; during the 1990s, seven such storms ravaged them. Now, more frequent El Niños (possibly spurred by global warming) seem to be bringing storms more often.

INCREASING FLOODS EXPECTED IN BANGLADESH

Bangladesh, one of the poorest countries on Earth, is likely to suffer disproportionately from global warming. Cyclones there historically kill many people; 130,000 people died in such a storm during April 1990. Less than one-fourth of Bangladesh's rural population has electricity; the country, as a whole, emits less than 0.1 percent of the world's greenhouse gases, compared to 24 percent by the United States (Huq 2001, 1617). Bangladesh is planning to use solar energy for new energy infrastructure but lacks the money to build sea walls to fend off rising sea levels.

Saleemul Huq, chairman of the Bangladesh Centre for Advanced Studies in Dhaka, Bangladesh, and director of the Climate Change Programme of the International Institute for Environment and Development in London, said that the world community has an obligation to pay serious attention to the views of people who stand to lose the most from climate change (Huq 2001, 1617).

A sea-level rise of half a meter (about twenty inches) could inundate about 10 percent of Bangladesh's habitable land, in 2004 the home of roughly 6 million people. A one-meter water-level rise would put 20 percent of the country (and 15 million people) under water (Radford 2004, 10). In addition to sea-level rise caused by warming, large parts of the Ganges Delta are subsiding because water has been withdrawn for agriculture, compounding the problem.

GLOBAL WARMING AMONG THE FACTORS DROWNING VENICE

Floods have plagued Venice for most of its history, but subsidence and slowly rising seas due to global warming have worsened flooding during the late twentieth and early twenty-first centuries. Venice, which sits atop several million wooden pillars pounded into marshy ground, has sunk by about 7.5 centimeters per century for the past 1,000 years. The rate is accelerating. In Venice, the water level rose to 125 centimeters above sea level in June 2002, a record for the month. At the beginning of the twentieth century, St. Mark's Square, the center of the city, was flooded an average of nine times a year. During 2001, it flooded almost 100 times. Venice flooded 111 times during 2003, more than any other year in its lengthy history. In another century, it will flood on a permanent basis ("Heavy Rains" 2002).

Venice has lost two-thirds of its population since 1950; the 60,000 remaining people host 12 million tourists a year, who make their way

Tourists browse a newspaper while seated on a flooded terrace facing San Marco Square, Venice, November 28, 1998. © Gerard Julien/AFP/Getty Images.

over planks into buildings with foundations rotted by perennial flooding. At the Danieli, one of Venice's most luxurious hotels, tourists arrive on wooden planks raised two feet above the marble floors, amid a suffocating stench from the high water (Poggioli 2002). Waters are rising around Venice for several reasons besides rising seas from global warming. During the twentieth century, mudflats that once impeded the sea's advance were dredged for shipping and other forms of development. Venice also is subsiding due to the removal of water from its aquifers for human use (Nosengo 2003, 609).

Venice residents and visitors have become accustomed to high-water drills for "acqua alta," high water. A system of sirens much like the ones that convey tornado warnings in the U.S. Midwest sounds when the water surges. Restaurants have stocked Wellington boots and moved their dining rooms upstairs. Venetian gondoliers ask their passengers to shift fore and aft—and to watch their heads—as they pass under bridges during episodes of high water (Rubin 2003). Some of the gondoliers have hacked off their boats' distinctive tailfins to clear the bridges that have been brought closer by rising waters.

Venice has proposed construction of massive retractable dikes in an attempt to hold the water at bay, amid considerable controversy. After seventeen years of heated debate, Venice's MOSE (*Modulo Sperimentale Elettromeccanico*) project will cost about U.S. $1 billion. Some environmentalists assert that the barriers will destroy the tidal movement required to keep local lagoon waters free of pollution and will thereby damage marine life. Water quality near Venice is already precarious, because pollution has leached into the lagoon from industry, homes, and motor traffic. The Italian Green Party favors shaping the lagoon's entrances to reduce the effects of tides, along with raising pavements as much as a meter inside Venice itself.

As proposed, the barriers will be constructed at the three entrances to Venice's natural lagoon from the Adriatic Sea. Each barrier is planned to house seventy-nine "flippers" that can be adjusted like the flaps of an aircraft. Installed below the water line, they will be raised when the sea level rises by more than one meter, which at the turn of the millennium was occurring about a dozen times a year (Watson 2001, 18).

During normal tides, according to an account in *Scotland on Sunday*, "the hollow barriers will sit within especially constructed trenches in the bed of the channels connecting the lagoon to the open sea. When a dangerously high tide is forecast, compressed air will be forced into the flippers which will have the effect of squeezing sea water out. As they rise, more water will trickle out to be replaced by air" (Watson 2001, 18). The project is expected to provide as many as 10,000 jobs during ten years of construction.

By mid-century, the flood-control (MOSE) system may be running almost all the time, severing the city from the ocean and transforming its neighboring lagoon, according to one observer, "into a stagnant pond with devastating effects on marine life and health" (Poggioli 2002). Many in Venice say that Project MOSE will not be of much help, because it will be constructed to operate only when water rises at least forty-three inches. Environmentalists have argued that the MOSE flood-control system is a construction boondoggle that will turn Venice into a toxic bathtub, in which the city's canals will be laced with sludge from surrounding heavy industry as well as the urban area's human waste. Venice's deputy mayor, Gianfranco Bettin, has called MOSE "expensive, hazardous, and probably useless" (Nosengo 2003, 608). Environmentalists have focused attention on bacteria from animal and human waste in the waters surrounding the city. "Venice has no sewage system; they just dump the stuff right out into the canals. It's not pretty," said Rick Gersberg, a microbiologist. "Normally, the tides come in and flush everything out. But when you cut off the tide, it just sits there" (Petrillo 2003).

14 GREENHOUSE GASES AND CHANGES IN THE OCEANS

In addition to the debate surrounding rising sea levels, a controversy has developed regarding whether changes in ocean circulation could cause some areas of the world (most notably, western Europe) to experience a marked cooling trend by suppressing the warm waters of the Gulf Stream, which help supply that area with a winter climate that is unusually warm for such a northern latitude. In addition, warming, especially in the equatorial Pacific Ocean, may be related to the frequency of El Niño episodes, with their dramatic effects on worldwide weather patterns. El Niño events bring increases in rainfall to some usually dry areas and debilitating drought to other regions. Warming ocean temperatures could lead to a slowing or cessation of the North Atlantic's thermohaline (overturning) circulation, which enriches deep waters with life-giving oxygen. Ocean circulation, in turn, is intimately related to the density of phytoplankton at the base of the oceanic food chain. Global warming could thus lead to widespread diminishment or extinction of many marine species.

WARMING AND INHIBITION OF THERMOHALINE CIRCULATION

The term "thermohaline" is used because "water density in the ocean is determined by both temperature and salinity" (Rahmstorf

2003, 699). According to Peter U. Clark and colleagues, writing in *Nature*, most, but not all, coupled general circulation model projections of twenty-first-century climate anticipate a reduction in the strength of the Atlantic overturning circulation, given warming seas provoked by increasing concentrations of greenhouse gases (Clark et al. 2002, 863–868). This flow is part of what marine scientists call the global conveyor, a vast submarine flow of water south from the Arctic. It is replaced by water flowing north from the tropics, the Gulf Stream, which keeps Britain 5 degrees C warmer than expected for its latitude. The Gulf Stream delivers 27,000 times more heat to British shores than do all of that nation's power stations (Radford [June 21] 2001, 3).

A report presented by Ruth Curry, a scientist at the Woods Hole Oceanographic Institute, at the annual meeting of the American Association for the Advancement of Science during February 2005 indicates that massive amounts of fresh water from melting Arctic ice are seeping into the Atlantic Ocean. According to Curry's research, between 1965

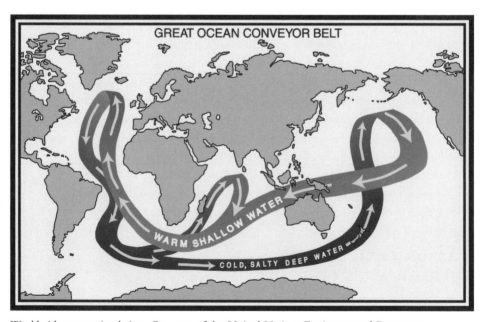

Worldwide ocean circulation. Courtesy of the United Nations Environmental Programme.

and 1995, about 4,800 cubic miles of fresh water (more water than Lake Superior, Lake Erie, Lake Ontario, and Lake Huron combined) melted from the Arctic region and poured into the northern Atlantic. Curry projected that if this pattern continues at current rates, the thermohaline circulation may begin to shut down in about two decades.

Curry, said that Greenland's ice, which to date has not been melting quickly, is now thawing at more rapid rates. "We are taking the first steps," Curry said in a news conference. "The system is moving in that direction" (Borenstein 2005, A-6). In the long term, according to calculations by Curry and Mauritzen, "at the observed rate, it would take about a century to accumulate enough fresh water (e.g., 9,000 cubic meters) to substantially affect the ocean exchanges across the Greenland-Scotland Ridge, and nearly two centuries of continuous dilution to stop them" (2005, 1774).

The salinity of the North Atlantic is intimately related to world ocean circulation. If the North Atlantic becomes too fresh (due, most likely, to melting Arctic ice), its waters could stop sinking, and the conveyor could slow, perhaps even stop. Spencer R. Weart, in *The Discovery of Global Warming* (2003), provided a capsule description of possible changes in the ocean's thermohaline circulation under the influence of sustained, substantial global warming: "If the North Atlantic around Iceland should become less salty—for example, if melting ice sheets diluted the upper ocean layer with fresh water—the surface layer would no longer be dense enough to sink. The entire circulation that drove cold water south along the bottom could lurch to a halt. Without the vast compensating drift of tropical waters northward, a new glacial period could begin" (p. 64).

Although atmospheric winds move heat quickly across large distances, fundamental changes in ocean temperatures and circulation may take centuries. Thermohaline circulation involves warm surface currents that distribute tropical heat, as well as deeper currents that carry cold water back toward the equator. Together, these currents form a system that circulates ocean water, heat, oxygen, and nutrients through the North and South Atlantic, then into the Indian Ocean and the Pacific. The North Atlantic is one of only two areas in the world's oceans where ice formation produces salty, dense water that then sinks and helps drive ocean circulation, carrying oxygen with it. The other area is the Weddell Sea in the Antarctic.

According to Peter U. Clark and colleagues, "If the warming is strong enough and sustained long enough, a complete collapse [of thermohaline circulation] cannot be excluded" (2002, 863). Clark further explained, "What is fairly clear is that if the ocean circulation patterns that now warm much of the North Atlantic were to slow or stop, the consequences could be quite severe. This might also happen much quicker than many people appreciate. At some point the question becomes how much risk do we want to take?" ("Slowing Ocean" 2002). The thermohaline circulation is vital to the mixture of oxygen and nutrients in much of the Atlantic Ocean and, thus, necessary to marine life as we know it. Similar circulation systems perform the same function in other oceans around the world. Without the thermohaline circulation, the oceans could become largely stagnant, with large areas bereft of life.

A report by the Environment News Service indicated that thermohaline circulation "is sensitive to moderate increases in temperature or influxes of fresh water. The cold, salty water that sinks in the far North Atlantic Ocean will not sink if it becomes a little bit warmer or a little bit less salty—and the change could happen in a matter of decades. . . . This system does not respond in what we call a linear manner. . . . Once you start putting on the brakes, this circulation pattern could slow down faster and faster and eventually stop altogether" ("Slowing Ocean" 2002). Developing themes first raised by Broecker (1987, 1997), Stocker and colleagues argue as well that these changes may be nonlinear: "They may have large amplitudes and may occur as surprises" (Stocker, Knutti, and Plattner 2001, 277). Inherent in such changes is their "reduced predictability." "Among such non-linear changes are the collapse of large Antarctic ice masses and rapid sea-level rise, the desertification of entire land regions, the thawing of permafrost and associated release of large amounts of radiatively active gases, and the collapse of the large-scale Atlantic Thermohaline (i.e., the temperature and salinity driven) circulation" (p. 277).

Changes in thermohaline circulation may have causes that are more complex than merely melting ice at northern latitudes. Knorr and Lohmann (2003, 532–536) argue that changes in the North Atlantic's circulation may originate far to the south. Using models, Knorr and Lohmann examined ways in which this circulation resumes after having stalled. "They find that, once a threshold is reached, slowly increasing sea-surface temperatures around Antarctica and receding ice cover lead to the North Atlantic Thermohaline circulation being rapidly switched on," wrote Thomas F. Stocker in *Nature* (2003, 496). This "switch" can warm parts of the North Atlantic by as much as 6 degrees C within a few decades. This view may represent something of a revolution in scientific understanding of how ocean circulation operates, or the models may be flawed—or some of each. Stocker has pointed out, for example, that Knorr and Lohmann's study ignores atmospheric dynamics, and, because of this limitation, "changes of ocean circulation patterns and their effect on the atmosphere-ocean heat exchange and the freshwater balance through the hydrological cycle cannot be accounted for" (2003, 499). According to Stocker, a solution to this dilemma awaits testing of more comprehensive models.

A study of North Atlantic ocean circulation published in late 2005 reported a 30 percent reduction in the Meridional Overturning (Thermohaline) Circulation at 25 degrees North latitude. Harry Bryden at the Southampton Oceanography Centre in the United Kingdom, whose group carried out the analysis, said he was not sure whether the change is temporary or part of a long-term trend. This analysis is, however, "the first observational evidence that such a decrease of the oceanic over-turning circulation is well underway" (Quadfasel 2005, 565).

"We don't want to say the circulation will shut down," Bryden told the *New Scientist*, "but we are nervous about our findings. They have come as quite a surprise" (Pearce 2005). Bryden's team measured north-south heat flow during 2004 using a set of instruments in various locations in the Atlantic between the Canary Islands and the Bahamas, then compared results to surveys in 1957, 1981, and 1992. They then calculated that the amount of water flowing north had fallen by around 30 percent (Bryden et al. 2005, 655). The area that was surveyed is limited, and thus could be influenced by regional variations.

PAST CHANGES IN THERMOHALINE CIRCULATION

Scientists studying past climate epochs have found times when the oceans' circulation did, indeed, lose its vital character, perhaps even shut down. Analyses of ice cores, deep-sea sediment cores, and other geologic evidence have clearly demonstrated that the conveyor has abruptly slowed or halted many times in Earth's past. That event has caused the North Atlantic region to cool significantly and brought long-term drought conditions to other areas of the Northern Hemisphere—over time spans as short as years to decades ("Study Reports" 2003). According to climate scientist Thomas F. Stocker, "Evidence from paleoclimatic archives suggests that the ocean atmosphere system has undergone dramatic and abrupt changes with widespread consequences in the past. Climatic changes are most pronounced in the North Atlantic region where annual mean temperatures can change by 10 degrees C. and more within a few decades. Climate models are capable of simulating some features of abrupt climate change. These same models also indicate that changes of this type may be triggered by global warming" (Stocker, Knutti, and Plattner 2001, 277).

At the end of the last ice age (between 8,200 and 12,800 years ago), ice-core records from Greenland indicate abrupt temperature declines at about the same time that massive amounts of cold fresh water were released from a huge body of glacial meltwater, which glaciologists today call Lake Agassiz. This lake extended from today's Canadian prairies eastward to Quebec and southward to Minnesota. Reaching the Atlantic Ocean through the St. Lawrence Valley and Hudson Bay, "such a large amount of low-density fresh water would have reduced the density of the North Atlantic surface water considerably, preventing it from sinking and thus slowing down (or perhaps even shutting off completely) the Gulf Stream," wrote glaciologist Doug Macdougall in *Frozen Earth: The Once and Future Story of Ice Ages* (2004, 110).

Evidence from the Irish Sea basin as recently as 19,000 years ago indicates a large reduction in the strength of North Atlantic deep-water formation that provoked cooling of the North Atlantic at the same time that global sea levels were rising rapidly, a response to melting ice. According to Clark and colleagues, "These responses identify mechanisms responsible for the propagation of deglacial climate signals to the Southern hemisphere and tropics while maintaining a cold climate in the Northern Hemisphere" (Clark et al. 2004, 1141).

Jochen Erbacher studied an anoxic event "in the restricted basins of the western Tethys and North Atlantic" during the mid-Cretaceous period, about 112 million years ago, which appears to have been caused by "increased Thermohaline stratification" (Erbacher et al. 2001, 325). The Western Tethys was a sea formed as continents were separating on the site of the present-day North Atlantic Ocean. "Ocean anoxic events were periods of high carbon burial that led to drawdown of atmospheric carbon dioxide, lowering of bottom-water oxygen concentrations and, in many cases, significant biological extinctions," commented Erbacher and colleagues (p. 325). "We suggest that the partial tectonic isolation of the various basins in the Tethys and Atlantic, a low sea level, and the initiation of warm global climates may be important factors in setting up oceanic stagnation" during this event (p. 327).

Seeking an analog to a world dominated by severe global warming, Paul A. Wilson and Richard D. Norris investigated the climate of the mid-Cretaceous period, "a time of unusually warm polar temperatures, repeated reef-drownings in the tropics, and a series of oceanic anoxic events, . . . with maximum sea-surface temperatures 3 to 5 degrees C. warmer than today" (Wilson and Norris 2001, 425). This period experienced considerable greenhouse forcing not unlike that expected by the end of the twenty-first century, except that the forcing was natural, not anthropogenic.

The possibility that water temperatures rose to 15 degrees C or perhaps higher in the Arctic during the Cretaceous period has also been explored by Hugh C. Jenkyns and colleagues in *Nature* (Jenkyns et al. 2004, 888–892). At the time, atmospheric carbon dioxide levels may have been three to six times today's levels, a "super greenhouse" caused mainly by volcanic out-gassing. Today's climate models have had a difficult time accommodating such warmth in the Arctic without raising projected temperatures at lower latitudes to levels that would have been intolerable for the animals and plants that lived at the time.

PRESENT-DAY EVIDENCE THAT THERMOHALINE CIRCULATION MAY BE BREAKING DOWN

Inhibition of global ocean circulation by warming is not a mere theory or a paleoclimatic curiosity. Evidence is accumulating that ocean circulation already is breaking down, although researchers are unsure whether this is evidence of a natural cycle, a provocation of warming sea waters, or both. Analysis has been restricted by the scanty nature of data prior to 1978. Evidence suggests that the North Atlantic has cooled, while the rest of the world has been warming—a probable sign of thermohaline disruption. During the last half of the twentieth century, research reports indicate a "dramatic" increase in fresh water released into the North Atlantic by melting ice. This "freshening" is well under way (Speth 2004, 61). According to scientists at the Woods Hole Oceanographic Institution, this is "the largest and most dramatic oceanic change ever measured in the era of modern instruments" (Gagosian 2003). By 2002, the amount of fresh water entering the Arctic Ocean was 7 percent more than during the 1930s (Speth 2004, 61).

According to oceanographer Ruth Curry, sea-surface waters in tropical regions have become dramatically saltier during the past fifty years, whereas surface waters at high latitudes, in Arctic regions, have become much fresher. These changes in salinity accelerated during the 1990s as global temperatures warmed. "This is the signature of increasing evaporation and precipitation" because of warming, Curry said, "and a sign of melting ice at the poles. These are consequences of global warming, either natural, human-caused or, more likely, both" (Cooke 2003, A-2). Richard A. Kerr, writing in *Science*, said that "to Curry and her colleagues, it's looking as if something has accelerated the world's cycle of evaporation and precipitation by 5 per cent to 10 per cent, and that something may well be global warming" (Kerr 2004,

35). These results indicate that fresh water has been lost from the low latitudes and added at high latitudes at a pace exceeding the ocean circulation's ability to compensate, the authors said ("Study Reports" 2003).

Curry led a team that examined salinity "on a long transect (50 degrees S. to 60 degrees North, or between Iceland in the north and the tip of South America) through the western basins of the Atlantic Ocean between the 1950s and the 1990s." They found "systematic freshening at both poleward ends contrasted with large increases of salinity pervading the upper water columns at lower latitudes." The authors asserted that their data extends "a growing body of evidence indicating that shifts in the oceanic distribution of fresh and saline waters are occurring worldwide in ways that suggest links to global warming and possible changes in the hydrologic cycle of the Earth" (Curry, Dickson, and Yashayaev 2003, 826).

This study suggests that relatively rapid oceanic changes and recent climate changes, including warming global temperatures, may be altering the fundamental planetary system that regulates evaporation and precipitation and that cycles fresh water around the globe. An acceleration of Earth's global hydrological cycle could affect global precipitation patterns, which govern the distribution, severity, and frequency of droughts, floods, and storms. The same change could exacerbate global warming by rapidly adding more water vapor—itself a potent, heat-trapping greenhouse gas—to the atmosphere. It could also continue to freshen North Atlantic Ocean waters to a point that could disrupt ocean circulation and trigger further climate changes ("Study Reports" 2003).

Japanese and Canadian scientists reported in *Nature* (Fukasawa et al. 2004, 825) that the deepest waters of the North Pacific Ocean have warmed significantly across the entire width of the ocean basin. Masao Fukasawa of the Japan Marine Science and Technology Center in Yokosuka, Japan, and five other researchers measured temperature changes by going to sea on three research vessels and measuring deepwater temperatures across the North Pacific. They then compared those temperature measurements to measurements made by researchers in

1985. The Fukasawa team's findings indicated a warming in the deep North Pacific over a "shorter time scale and larger spatial scale than [has] ever been believed. As [far] as I know, our result is the first which shows such a large-scale temperature change in the global Thermohaline circulation," that is, in global heat and salt circulation via ocean currents (Davidson 2004, A-4). It's much too soon to blame global warming for the deep-sea warming, Reid and other experts cautioned. "To go from this [observation] to say, 'This is global warming,' is just a guess," Reid said. In any case, "this [Fukasawa result] is something that should be watched very carefully" (p. A-4).

Michael J. McPhaden and Dongxiao Zhang, writing in *Nature* (2002), also made a case that thermohaline circulation is slowing:

> Some theories ascribe a central role [in global climate change] to the wind-driven meridional overturning circulation between the tropical and subtropical oceans. Here we show, from observations over the last 50 years, that this overturning circulation has been slowing down since the 1970s, causing a decrease in upwelling of about 25 per cent in an equatorial strip between 9 degrees north and 9 degrees south. This reduction in equatorial upwelling of relatively cool water . . . is associated with a rise in equatorial sea-surface temperatures of about 0.8 degrees C. (p. 603)

Paleoclimatic records indicate that once climate change is "tripped," change may be quite rapid. For example, various proxy records from sources such as ice cores indicate temperature changes of up to 16 degrees C in Greenland in a decade (Stocker, Knutti, and Plattner 2001, 289). Could such a rapid change, triggered by global warming, shut down the oceans' circulation system within a matter of decades? Terrence Joyce, chairman of the Physical-Oceanography Department at Woods Hole Oceanographic Institution in Massachusetts, has been trying to raise awareness about this possibility. He says he is "not predicting an imminent climate change—only that once it started (and it is getting more likely) it could occur within 10 years" (Cowen 2002, 14). Woods Hole director Robert Gagosian feels an urgency to settle the question. He sees enough disturbing information in the North Atlantic data that oceanographers from Woods Hole and other institutions have gathered to elicit "strong evidence that we may be approaching a dangerous threshold" (p. 14).

Stocker and colleagues have written: "Although the magnitude of the change is highly uncertain, the models agree that the Thermohaline circulation in the Atlantic will reduce due to the gain of buoyancy associated with the warming and a stronger hydrological cycle" (Stocker, Knutti, and Plattner 2001, 290). R. B. Thorpe and others consider a number of model variations indicating that various degrees of global warming may decrease thermohaline circulation in the Atlantic between 20 and 60 percent due to temperature change (greater warming at high latitudes) and 40 percent due to salinity change (Thorpe et al. 2001, 3102).

B. Dickson and colleagues wrote in *Nature* that "from observations it has not been possible to detect whether the ocean's overturning circulation is changing, but recent evidence suggests that the transport over the sills [in the ocean] may be slackening. Here we show, through the analysis of hydrographic records, that the system of overflow and entrainment that ventilates the deep Atlantic has steadily changed over the past four decades. We find that these changes have already led to sustained freshening of the deep ocean" (Dickson et al. 2002, 832).

Dickson and colleagues' ideas have received support from other studies. An international team of hydrologists and oceanographers reported in the December 13, 2002, edition of *Science* that the flow of fresh water from Arctic rivers into the Arctic Ocean has increased significantly over recent decades. If the trend continues, some scientists believe it could impact the global climate, perhaps leading to cooling in northern Europe ("Study Reveals" 2002).

Bruce J. Peterson of the Marine Biological Laboratory's Ecosystems Center led a research team of scientists from the United States, Russia, and Germany whose members analyzed discharge data from the six largest Eurasian rivers that drain into the Arctic Ocean. These rivers, all with headwaters in Russia, account for more than 40 percent of the total freshwater inflows into the Arctic Ocean. Peterson and colleagues found that combined annual discharge from the Russian rivers increased by 7 percent from 1936 to 1999. They contend that this measured increase in runoff is an observed confirmation of what climatologists have been saying for years—that freshwater flow to the Arctic Ocean and North Atlantic will increase with global warming ("Study Reveals" 2002).

"If the observed positive relationship between global temperature and river discharge continues into the future, Arctic river discharge may increase to levels that impact Atlantic Ocean circulation and climate within the 21st century," said Peterson. Annual discharge of the rivers into the Arctic increased about 128 cubic kilometers per year during the sixty-three-year study period. The authors of the study warned that increasing river discharge, coupled with icemelt from Greenland, could have important effects on thermohaline circulation during the twenty-first century (Peterson et al. 2002, 2171–2172; "Study Reveals" 2002).

In another indication that the thermohaline circulation may be slowing, Peter Wadhams of the Scott Polar Research Institute reported that, for the fifth time in seven years, a huge tongue of shelf ice had failed to form in the Greenland Sea (Radford [June 21] 2001, 3). This ice shelf is a powerful driver of thermohaline circulation. During 1994, Wadhams told a major oceanography conference that convection in the Greenland Sea had "virtually stopped in the last decade" (Leggett 2001, 150). Ten years earlier, Wadhams had measured convection in the Greenland Sea all the way to the bottom (3,500 meters). Five years later, convection was limited to the top 2,000 meters of water. By 1993, mixing was restricted to 1,000 meters, and, in 1994, it ceased altogether. "It is a process which has been going on for several thousand years," Wadhams said. "And when you see it decline sharply over a decade it gets you worried" (p. 150).

During mid-2001, other scientists also disclosed evidence that the thermohaline circulation is, indeed, breaking down. Researchers in Denmark's Faeroe Islands, halfway between Iceland and the northern tip of Britain, found evidence of a 20 percent drop since 1950 in the volume of deep cold water flowing south from the Arctic region through one of several channels into the North Atlantic. Most of the decrease has occurred during the past thirty years, and the rate of decline has accelerated in the past five years. This study, reported in *Nature*, combined water-flow readings from the ocean floor after 1995 with fifty years of temperature and salinity records from weather ships in the area.

Additional proof of thermohaline breakdown is evident from a rising volume of fresh water in the Norwegian Sea from melting sea ice. "The seas are warmer and there is more freshwater not just from melting sea ice but from the Siberian rivers. The water has to be salty and dense or it just won't sink," researcher Bogi Hansen said (Connor 2001, 14). "Estimating the volume flux conservatively, we find a decrease by at least 20 per cent relative to 1950. If this reduction in deep flow from the Nordic seas is not compensated by increased flow from other sources," wrote Hansen and colleagues (2001, 927), "it implies a weakened global Thermohaline circulation and reduced inflow of Atlantic water to the Nordic seas."

The findings most likely indicate an equal drop in the amounts of warm surface water flowing north through the same passage, said Hansen. "The motor driving the ocean conveyer belt appears to be slowing down" (Calamai 2001, A-18; Hansen, Turrell, and Sterhus 2001, 927). According to a report by Steve Connor in the *London Independent*, "The research centered on measurements taken of water movements in a deep-sea channel that separates the Faeroes from northern Scotland. The current in the Faeroe Bank channel pushes about 2 million cubic meters of water every second into the Atlantic, an amount equivalent to about twice the total flow of all the rivers of the world combined" (Connor 2001, 14). "If this reduction we have seen in the Faeroe Bank channel is also seen in the Denmark Strait then we can be sure that the North Atlantic flow has been reduced," Hansen said (Connor 2001, 14). Commenting on the Hansen study, Andrew Weaver, who holds the Canada Research Chair in Atmospheric Studies at the University of Victoria, said, "This is yet another very important brick in a very solid wall of evidence about the reality of global climate change" (Calamai 2001, A-18).

Robert C. Cowen, writing in the *Christian Science Monitor*, summarized the state of the research in the early twenty-first century: "Water-circulation data from the North Atlantic now suggest the climate system may be approaching a threshold; if man-made warming or natural causes push it over the edge, the system will chill down many temperate parts of North America and Europe, even while the planet as a whole continues to warm" (2002, 14).

Many existing climate models do not factor in the possibility that the North Atlantic circulation pump could shut down. A failure of the ocean's circulation system may not only imperil sea life but could accelerate climate change in some nasty and surprising ways. In Gagosian's words, it could "freeze rivers and harbors and bind North Atlantic shipping lanes in ice, ... disrupt the operation of ground and air transportation, ... cause energy needs to soar exponentially, ... [and] force wholesale changes in agricultural practices and fisheries. Efforts to curb carbon-dioxide emissions to slow global warming would become a secondary issue as people tried to cope with more immediate challenges" (p. 14).

THERMOHALINE CIRCULATION:
DEBATING POINTS

The idea that the failure of thermohaline circulation could cause colder temperatures in land masses around the North Atlantic Ocean is highly debatable, as well as politically charged. The debate was given extra force in 2001 when one of the idea's major proponents backed away from it. Wallace S. Broecker, of Columbia University, theretofore advocating that the "business as usual" fossil fuel track "[would] run the risk, late in the twenty-first century, of triggering an abrupt reorganization of the Earth's Thermohaline circulation" (Broecker 2001, 83). He basically repudiated his earlier position:

The recent discovery by Gerard Bond that the 1,500-year cycle that paced these glacial disruptions continued in a muted form during times of interglaciation casts a new light on this situation. . . . It leads me to suspect that the large and rapid atmospheric changes of glacial time were driven by a sea-ice amplifier. If so, then, because little sea ice will remain at the time of a greenhouse-induced Thermohaline reorganization, perhaps the threat will be far smaller than I had previously envisioned. (Broecker 2001, 83)

Broecker said, "I apologize for my previous sins" of overemphasizing the Gulf Stream's role (Kerr 2002, 2202).

As is so often the case regarding global warming's putative effects on the Earth's ecosystem, the premises under examination are open to dispute. Is the Gulf Stream *really* the main climate driver warming Europe's winters? In October 2002, a team of scientists writing in the *Quarterly Journal of the Royal Meteorological Society* asserted that this popular assumption is incorrect. Rather, they maintained, Europe is warmed "by atmospheric circulation tweaked by the Rocky Mountains [and] . . . summer's warmth lingering in the North Atlantic" (Kerr 2002, 2202).

Richard Seager of Columbia University's Lamont-Doherty Earth Observatory in Palisades, New York, and David Battisti of the University of Washington headed this study, which sought to determine the relative influences of various factors on European climate. They noted that winds carry five times as much heat out of the tropics to the mid-latitudes than do oceanic currents. They also estimated that roughly "80 per cent of the heat that cross-Atlantic winds picked up was summer heat briefly stored in the ocean rather than heat carried by the Gulf Stream" (Kerr 2002, 2202). Seager and colleagues relegate the Gulf Stream to the role of a minor player in Europe's wintertime climate. They asserted, however, that the Gulf Stream *does* play a major role in warming Scandinavia and in keeping the far northern Atlantic free of ice (Seager et al. 2002, 2563).

OCEAN CIRCULATION AND FREQUENCY OF EL NIÑO EVENTS

The most definitive support for the idea that a warmer Earth promotes El Niño conditions was published in *Science* during the summer of 2005. Michael W. Wara and colleagues reported that during the warm early Pliocene, between 3 and 4.5 million years ago (the most recent interval in global climate that was markedly warmer than the present), El Niño—type conditions were predominant in the Pacific Ocean. "Sustained El Niño-type conditions . . . could be a consequence of, and play an important role in determining, global warmth," they wrote (Wara, Ravelo, and Delaney 2005, 758). This study contradicted another, compiled from much less data, that made a case for a persistent cool La Niña state in the eastern Pacific during the early to mid-Pliocene (Kerr 2005, 687; Rickaby and Halloran 2005, 1948–1952).

Writing in *Geophysical Research Letters* during 1996, Kevin E. Trenberth and Timothy J. Hoar first suggested that El Niño cycles and global warming could be related (p. 57). Other researchers have agreed with this idea. "A shift towards stronger and more frequent El Niño events occurred in 1976–77, just when the decadal decline in the [thermohaline] circulation began," Heike Langenberg, an oceanographer, wrote in *Nature*. "But it is not clear whether the longer-term variation in circulation is a consequence of the increasing number of El Niño years, or whether the circulation slowdown provided the background for the change" (Connor [February 7] 2002). Robert Dunbar of Stanford University, writing in *Nature*, also made a case that the change in frequency of El Niño events has coincided with a shift in climate since 1976 and 1977. "The northern and tropical Pacific warmed abruptly and stayed warm for the next two decades," he wrote (Nuttall [October 26] 2000). Dunbar suspects that the more frequent El Niños are being triggered by higher sea-surface temperatures since the late 1970s, which in turn have been linked to the buildup of greenhouse gases. Dunbar contends that further change in El Niño is likely as more carbon dioxide is pumped into the atmosphere.

El Niño events became more intense and frequent around 1976, about the same time that some scientists have documented a slowing in thermohaline circulation. Before 1976, El Niño events averaged once every seven years; since then, they are occurring roughly every four years. "There was a period in the early 1990s when three weak El Niños occurred in fairly rapid succession," said oceanographer Michael

McPhaden. "Over the past 25 years, there has been a higher frequency of El Niños which were stronger and longer lasting" than usual (Cooke 2002, A-46; McPhaden and Zhang 2002, 603). During the 1990s, for example, El Niño conditions occurred twice (1990–1994 and 1997–1998), whereas, in the cooler ocean waters of the mid-nineteenth century, the typical pattern was one El Niño every ten to fifteen years. The coral record also indicates that ocean warming is not steady. It occurs in surges, the most notable of which, recently, was during 1976–1977, predating the increase in El Niño episodes.

The El Niño pattern causes dramatic warming of equatorial waters adjacent to the coast of South America, an area that is usually cooler than surrounding waters. It can affect air-circulation patterns (and therefore weather) in North and South America, as well as across the Pacific Ocean into southern Asia. Most notably, these shifting patterns can turn the western coast of South America, which is usually a desert, into an area plagued by torrential rains, while changing the Amazon Valley, which is usually wet, into a much drier region. This change takes place as the direction of winds across the Andes changes from generally east-to-west to west-to-east. The El Niño pattern also restricts the supply of moisture to Asia, hobbling the usual monsoon pattern, with widespread effects for hundreds of millions of people from India, through Indonesia, to Australia.

Mat Collins, a senior research fellow at the Meteorological Office in Reading, England, has raised a possibility that persistent El Niño conditions accelerated by global warming could change prevailing winds in South America in such a way that the Amazon Valley may become a desert by the end of the twenty-first century. "In our model, 50 years from now the Amazon dries up and dies," he told the British Association at the University of Salford (Arthur [September 9] 2003, 6).

If the Amazon Valley dries up, Arthur said that it may compound global warming: "There would be a reinforcing effect because, as the rainforest dried up, the carbon that is presently locked in its vegetation would be released into the atmosphere" ([September 9] 2003, 6). Collins presently rates the chance of such a scenario at between 10 and 20 percent, what he calls a preliminary estimate. Desert conditions in the Amazon could be initiated, said Collins, by a "super El Niño" that would cause prevailing winds in equatorial South America to surge over the Andes from the west, causing floods there, as is usual during El Niño episodes. Collins believes that warming will increase the number, intensity, and duration of El Niño events, which bring desiccating westerly winds to tumble down the Andes into the Amazon Valley, reversing the usual up-slope flow that helps cause copious rainfall.

Alexander W. Tudhope and colleagues' studies "provide strong support for the idea that ENSO [El Niño Southern Oscillation, El Niño's technical name] may be more responsive to global change than previously thought" (Cole 2001, 1496). The two most pronounced ENSO episodes in recorded weather history have occurred since 1980, in conjunction with rapidly rising global temperatures, leading to speculation that warmer weather may be associated with more frequent and intense El Niño episodes. Tudhope and colleagues found evidence that El Niño episodes have occurred for at least 130,000 years, under both glacial and interglacial conditions, but that ENSO episodes usually increase with warmer weather. This study used coral populations off the New Guinea coast as a climate proxy. Rapid tectonic lift in the area has exposed corals that are much older than most available samples. "At their site," reported Julia Cole in *Science*, "modern (late nineteenth century to today) corals show the highest amplitude of inter-annual ENSO variance of all samples over the last 130,000 years" (p. 1497).

The relationship between warming and more frequent El Niños is complicated by other factors, however. Variations in the Earth's orbit (and changing intensities of incoming sunlight) also may play a role. According to Michael J. McPhaden and Dongxiao Zhang, writing in *Nature*:

Results of some numerical coupled ocean-atmosphere model studies suggest that the response of the tropical Pacific to greenhouse-gas forcing resembles a permanent El Niño-like condition. Thus, it is conceivable that climate fluctuations in the tropical Pacific over the past 50 years, including the recent slowdown in meridional overturning circulation, may have been influenced by global warming as well as by natural variability. Unfortunately, separating out the putative effects of anthropogenically forced climate change from natural variations is not possible with relatively short data records. (2002, 603)

Measurements collected by McPhaden and Zhang indicate that the flow pattern that carries warm tropical surface waters to the northern Pacific Ocean and the return flow that takes cooler water at depth toward the equator "has been slowing down since the 1970s" (Cooke 2002, A-46; McPhaden and Zhang 2002, 603). As for the sea-surface circulation, McPhaden, who works at NOAA's Pacific Marine Environmental Laboratory in Seattle, and Zhang, at the University of Washington, contend that the slowing of sea-water transport could turn out to be signs of global change. The newly analyzed data show that, in addition to El Niño, "there's something else out there elevating temperatures." As a result, "when you have an El Niño, you get sort of a double dose; it just accentuates an El Niño.... It is conceivable that climate fluctuations in the tropical Pacific may have been influenced by global warming, as well as by natural variability." It is not yet possible, however, to sort out whether, or how much of, the change is due to the release of greenhouse gases from the burning of fossil fuels (Cooke 2002, A-46).

Frank E. Urban and colleagues writing in *Nature* (2000) described the use of oxygen-isotope ratios in the carbonate skeletons of reef corals from Maiana Atoll in the central Pacific in their attempt to chart sea-surface temperatures and salinity levels from 1840. Such charting could extend instrument readings that began only during the late 1950s. The Urban team's findings indicate that the length of El Niño cycles varies rather dramatically "with even small changes in tropical temperatures" (p. 989).

Matthew Huber and Rodrigo Caballero looked at the warm climate of the Eocene, 55–35 million years ago, as a testing ground for an interaction between a "hothouse" climate and frequency of El Niño cycles. Comparing climate simulations of that period with variability records preserved in lake sediments, they found little connection. "The simulations show Pacific deep-ocean and high-latitude surface warming of about 10 degrees C. but little change in the tropical thermocline structure, atmosphere-ocean dynamics, and ENSO [El Niño], in agreement with proxies," they wrote in *Science*. "This result contrasts with theories linking past and future 'hothouse' climates with a shift toward a permanent El Niño-like state" (2003, 877).

15 GLOBAL WARMING AND MARINE LIFE

Warming seas have begun to change patterns of life in the oceans in fundamental ways, as rising carbon dioxide levels in ocean water threaten some species of sea life. At the turn of the millennium, scientists were already beginning to notice an alteration in the acidity of the oceans due to their carbon dioxide levels. Though the oceans have long been thought to be "sinks," or repositories, of humankind's excess carbon dioxide, their capacity is not limitless. By the turn of the millennium, with carbon dioxide levels in the oceans probably higher than they have been at any time since the days of the dinosaurs, scientists were discovering just how human alterations of the atmosphere's composition also could fundamentally change the makeup of the oceans and the life they harbor.

Free-floating, microscopic plants called phytoplankton, the base of the oceanic food web, have been declining rapidly in areas where they once nourished rich fisheries, most notably in the temperate areas of the world's oceans in the Northern and Southern hemispheres. Cod and salmon, both cold-water fish, are becoming scarcer in English offshore waters, as tropical species are being sighted in the same areas more frequently. Sharks have been sighted in Alaskan waters.

Around the world, coral reefs have suffered due to warming above their tolerances in tropical oceans. Not all sea life has suffered because of warming in the oceans, however. Jellyfish, for example, thrive in warming waters. They also benefit from some forms of human water pollution, which has led to increasing size and toxicity of their stings. Giant squid also seem to benefit from warming seas.

RISING CARBON DIOXIDE LEVELS IN THE OCEANS MAY IMPERIL MARINE LIFE

As the twenty-first century dawned, carbon dioxide levels were rising in the oceans more rapidly than at any time since the age of the dinosaurs, according to a report published September 25, 2003, in *Nature*. Ken Caldeira and Michael E. Wickett wrote: "We find that oceanic absorption of CO_2 from fossil fuels may result in larger pH changes over the next several centuries than any inferred in the geological record [during the last] 300 million years, with the possible exception of those resulting from rare, extreme events such as bolide impacts or catastrophic methane hydrate degassing" (2003, 365). A "bolide" is a large extraterrestrial body (usually at least a half mile in diameter, perhaps much larger) that impacts the Earth at a speed roughly equal to that of a bullet in flight.

Rising carbon dioxide levels in the oceans could threaten the health of many marine organisms. Regarding the acidification of the oceans, "we're taking a huge risk," said Ulf Riebesell, a marine biologist at the Leibniz Institute of Marine Sciences in Kiel, Germany. "Chemical ocean conditions 100 years from now will probably have no equivalent in the geological past, and key organisms may have no mechanisms to adapt to the change (Schiermeier 2004, 820). "If we continue down the path we are going, we will produce changes greater than any experienced in the past 300 million years—with the possible exception of rare, extreme events such as comet impacts," Caldeira of the Lawrence Livermore National Laboratory warned (Toner 2003). Since carbon dioxide levels began to be measured on a systemic basis worldwide in 1958, its concentration in the atmosphere has risen 17 percent.

Until now, some climate experts have asserted that the oceans would help to control the rise in carbon dioxide by acting as a filter. Ken Caldeira and Michael Wickett said, however, that carbon dioxide that is removed from the atmosphere enters the oceans as carbonic acid, gradually altering the acidity of ocean water. According to their studies, the change over the last century already has fundamentally altered the chemistry of the oceans. Caldeira pointed to acid rain from industrial emissions as a possible precursor of changes in the oceans. "Most ocean life resides near the surface, where the greatest change would be expected to come, but deep ocean life may prove to be even more sensitive to changes," Caldeira asserted (Toner 2003).

Marine plankton and organisms whose skeletons or shells contain calcium carbonate, which can be dissolved by acid solutions, may be particularly vulnerable. Coral reefs, which already are suffering from pollution, rising ocean temperatures, and other stresses, are comprised almost entirely of calcium carbonate. "It's difficult to predict what will happen because we haven't really studied the range of impacts," Caldeira said. "But we can say that if we continue business as usual, we are going to see some significant changes in the acidity of the world's oceans" (Toner 2003).

Additional concern that human-generated carbon dioxide may be acidifying the oceans was expressed by a scientific team headed by Richard A. Feely, a scientist with the National Oceanic and Atmospheric Administration's Pacific Marine Environmental Laboratory in Seattle, and Christopher L. Sabine, also of the NOAA laboratory. Although the oceans have reached only one-third of their capacity for absorbing humankind's excess carbon dioxide, the rising level of carbon in ocean water (and, therefore, its acidity) may be impeding sea animals' ability to grow protective shells (Feely et al. 2004, 362; "Report Says" 2004; Freely et al. 2004, 367).

Warming-provoked acidification of the oceans has a precedent in the Earth's natural history, notably during the Paleocene-Eocene Thermal Maximum, when roughly 2 trillion metric tons of carbon, oxidized from methane clathrates, surged into the atmosphere from the oceans, raising sea-surface temperatures by 5 degrees C in the tropics and up to about 9 degrees C at high latitudes. An initial, rapid rise in temperatures over 1,000 years was followed by a slower increase during the next 30,000 years (Zachos et al. 2005, 1612).

Scientists have been studying the acidification of the oceans during this long-ago epoch as an analog to similar conditions anticipated in response to human-provoked increases in carbon dioxide. One study of the problem, published in *Nature*, pointedly concluded: "What, if any, implications might this have for the future? If combustion of the entire fossil-fuel reservoir (about 4,500 gigatons of carbon) is assumed, the impacts on deep-sea pH [acidity] and biota will likely be similar to those in the Paleocene-Eocene Thermal Maximum. However, because the anthropogenic carbon input will occur within just 300 years, which is less than the mixing time of the ocean, the impacts on ocean surface pH and biota will probably be more severe" (Zachos et al. 2005, 1614).

By 2005, the date at which increasing carbon dioxide levels in the oceans were expected to change acidity enough to dissolve the calcium-carbonate shells of corals, planktons, and other marine animals was being advanced to the next few decades, sooner than previously projected. A team of scientists writing in *Nature* said that "In our projections, Southern Ocean surface waters will begin to become under-saturated with respect to aragonite, a metastable form of calcium carbonate, by the year 2050. By 2100, this under-saturation could extend throughout the entire Southern Ocean and into the subarctic Pacific Ocean" (Orr et al. 2005, 681).

PHYTOPLANKTON DEPLETION AND WARMING SEAS

Gregory Beaugrand and colleagues reported in *Science* that several sea species, notably zooplankton (a major base of the oceanic food chain), moved northward between 1960 and 1999 in response to warming water temperatures: "We provide evidence of large-scale changes in the bio-geography of calanoid copepod crustaceans in the Eastern North Atlantic Ocean and European shelf seas. . . . Strong bio-geographical shifts in all copepod assemblages have occurred with a northward extension of more than 10 degrees latitude of warm-water species associated with a decrease in the number of colder-water species" (2002, 1692).

This study was based on the analysis of 176,778 samples collected by the Continuous Plankton Recorder Survey taken monthly in the North Atlantic since 1946. The scientists wrote: "The observed bio-geographical shifts may have serious consequences for exploited resources in the North Sea, especially fisheries. If these changes continue, they could lead to substantial modifications in the abundance of fish, with a decline or even a collapse in the stock of boreal species such as cod, which is already weakened by over-fishing" (Beaugrand et al. 2002, 1693–1694).

Mike Toner wrote in the *Atlanta Journal-Constitution*: "Plankton are also as important to the long-term health of the atmosphere as the world's forests. The photosynthesis of the ocean's tiny green plants account for about half of the carbon dioxide that plants remove from the atmosphere each year" (Toner 2002, 3-A). "The less phytoplankton you have, the less carbon is taken up by the oceans," said Margarita Conkright, of the National Oceanic and Atmospheric Administration (p. 3-A).

While studying nitrogen balances, Canadian and American oceanographers found evidence that warmer oceans will lose nitrate, depleting phytoplankton and boosting atmospheric carbon dioxide levels (Leggett 2001, 226). Satellite surveys have detected a sharp decline in plankton in several of the world's oceans, a potential threat to the marine food chain and a result that could undercut one of the world's natural buffers to global warming (Toner 2002, 3-A). Phytoplankon in the world's oceans are a key "carbon sink," but warming of the oceans may be devastating them. Surveys by satellites and ships have confirmed that the productivity of these microscopic plants is declining, most notably from the North Pacific to the high Arctic.

The severity of phytoplankton's decline varies from ocean to ocean, scientists reported during the summer of 2002. The greatest decline has been in the North Pacific Ocean, where summer levels have dropped by more than 30 percent since the 1980s. "It's difficult to say what the implications are, but they could be pretty significant," said Watson Gregg of NASA's Goddard Space Flight Center. "The whole marine food chain depends on the health and productivity of phytoplankton" (Toner 2002, 3-A).

Comparing sets of satellite data from early 1980 to the late 1990s, these researchers reported in the August 8, 2002, issue of *Geophysical Research Letters* that sharp declines in plankton had taken place in *both* the North Pacific and in the North Atlantic, where their abundance decreased by 14 percent. In equatorial regions, plankton levels increased. Worldwide, plankton stocks decreased more than 8 percent (Gregg and Conkright 2002).

The researchers were not certain whether the decline of phyto-plankton is part of a natural cycle in the oceans, a reflection of regional changes, or a result of a gradual warming of the Earth. They did, however, find a close correlation between the decline of plankton and increasing ocean surface temperatures, indicating that "climate change could be a cause as well as an effect of plankton declines" (Toner 2002, 3-A). Plankton need two things to grow: sunlight and nutrients. The researchers said warmer sea-surface temperatures interfere with up-welling of colder water that is rich in nutrients from the oceans' depths.

Scientists in Australia have found similar declines in phytoplankton populations and have asserted that global warming is starving the depths of the Southern Ocean of oxygen. Research scientists working with the Australian climate-research agency CSIRO said they have found a significant drop in the Southern Ocean's oxygen levels during the past thirty years; they also anticipate that this situation will worsen in the future. Hobart-based scientist Richard Matear said that oxygen-starved oceans may lead to long-term devastation of marine life. Matear said that the decline in oxygen levels was most likely caused by global warming, as natural variability could not explain the reduction. "The interpretation is that less oxygen-rich water is penetrating into the ocean and this in turn gives additional credibility to climate change models." Matear commented that the Southern Ocean was considered by oceanographers to be the "lungs" of the world's oceans, creating 55 percent of the water that regenerates the deep ocean. Any decrease in the Southern Ocean's capacity could have consequences for oceans around the world (Barbeliuk 2002).

Matear asserted that the Southern Ocean itself is unlikely to be greatly impacted by the change, since it is so oxygen-rich. Other oceans that already have limited oxygen, however, could suffer dire consequences, with the variety of marine life greatly reduced. "Most organisms that live in the ocean require oxygen, only a very few don't," Matear said (Barbeliuk 2002). Warmer waters are unable to carry as much oxygen or to sink as deeply into the ocean's depths. The colder the water, the faster it sinks, acting as an oxygen pump. The Southern Ocean's present oxygen level is about 200 micromoles per kilogram, a decline of about 15 micromoles during the thirty years leading up to the year 2000. Matear and fellow scientists Tony Hirst, also of the CSIRO, and Ben McNeil, from the Antarctic Co-operative Research Centre, used chemical data gathered during oceanographic research voyages south of Australia to look for changes in the ocean conditions. "Having demonstrated that oxygen is a valuable indicator of climate change in our models, we now have a quantity to monitor to detect future changes," he said (Barbeliuk 2002).

The decline of plankton also is related to the collapse of the thermohaline circulation (the Atlantic meridional overturning circulation, described in Chapter 14). Writing in *Nature*, Andreas Schmittner reported on results of models simulating disruption of the Atlantic meridional overturning circulation that lead to a collapse of the North Atlantic plankton stocks to less than half of their initial biomass, "owing to rapid shoaling of winter mixed layers and their associated separation from the deep ocean nutrient reservoir." Schmittner wrote that "these model results are consistent with the available high-resolution palaeorecord, and suggest that global ocean productivity is sensitive to changes in the Atlantic meridional overturning circulation" (Schmittner 2005, 628).

Adding more support to the idea that phytoplankton populations are declining around the world because of deteriorating ocean circulation, British scientists examining the Atlantic Ocean south of Iceland found that populations of zooplankton, which feed many larger ocean species, have declined by as much as 90 percent in four decades. This may portend population reductions for larger species, from cod and haddock to whales and dolphins. "This is deeply worrying," said marine biologist Phil Williamson of East Anglia University. "We don't know why zooplankton numbers have plummeted, though global warming looks [like] the best candidate. What is certain is that removing the bottom link from the ocean food chain could have profound and unpleasant results" (Knutti et al. 2004, 851–854; McKie 2001, 14).

Updating a major survey of zooplankton levels in the North Atlantic completed during 1963, a team of British scientists during November of 2001 set out in the marine research vessel *Discovery* to measure changes in these levels. Using automated equipment, the scientists sampled concentrations of *Calanus finmarhicus*, the principal type of Atlantic zooplankton. Having carried out 800 samplings in an area 1,000 miles south of Iceland, the scientists found 5,000 to 10,000 zooplankton per square meter instead of the 50,000 average found in the earlier survey (McKie 2001, 14). The scientists believe that gradually increasing sea temperatures are playing a major role in the zooplankton's decline.

Whether decline of plankton is directly related to increased ocean temperatures is not yet certain, according to Watson W. Gregg, a NASA biologist at the Goddard Space Flight Center in Greenbelt, Maryland, because other factors also affect the productivity of phytoplankton, such as the availability of iron. According to Gregg, the greatest loss of phytoplankton has occurred where ocean temperatures have risen most significantly between the early 1980s and the late 1990s. In the North Atlantic, during summer, sea-surface temperatures rose about 1.3 degrees F during that period, Gregg said, while in the North Pacific the ocean's surface temperatures rose about 0.7 degree F. "This research shows that ocean primary productivity is declining, and it may be the result of climate changes such as increased temperatures and decreased iron deposition into parts of the oceans," Gregg said. "This has major implications for the global carbon cycle" (Perlman 2003, A-6).

Warming may have aided phytoplankton growth in some areas (unlike northern latitudes of the Atlantic Ocean, where it is declining with rising temperatures). Declining winter and spring snow cover over Eurasia has contributed to an increasing land/ocean temperature gradient that enhances summertime monsoon winds over the western Arabian Sea, near the coasts of Somalia, Yemen, and Oman. According to one study, increased upwelling in the area contributes to an increase of more than 350 percent in average summertime phytoplankton biomass along the coast and 300 percent offshore, making the area more productive of aquatic life (Goes et al. 2005, 545).

ECOLOGICAL "MELTDOWN" IN THE NORTH SEA

Global warming is contributing to changes that some scientists describe as an "ecological meltdown," with devastating implications for fisheries and wildlife. The "meltdown" begins at the base of the food chain, as increasing sea temperatures reduce plankton populations, described previously. The devastation of the plankton then ripples up the food chain as fish stocks and sea-bird populations decline as well.

Scientists at the Sir Alistair Hardy Foundation for Ocean Science in Plymouth, England, which has been monitoring plankton growth in the North Sea for more than seventy years, have said that unprecedented warming of the North Sea has driven plankton hundreds of miles to the north. They have been replaced by smaller, warm-water species that are less nutritious (Sadler and Lean 2003, 12). Overfishing of cod and other species has played a role, but fish stocks have not recovered after cuts in fishing quotas. The number of salmon returning to British waters is now half what it was twenty years previously. A decline in plankton stocks is a major factor in this decrease.

"A regime shift has taken place and the whole ecology of the North Sea has changed quite dramatically," said Chris Reid, the foundation's director. "We are seeing a collapse in the system as we knew it. Catches of salmon and cod are already down and we are getting smaller fish. We are seeing visual evidence of climate change on a large-scale ecosystem. We are likely to see even greater warming, with temperatures becoming more like those off the Atlantic coast of Spain or further south, bringing a complete change of ecology" (Sadler and Lean 2003, 12).

According to a report by Richard Sadler and Geoffrey Lean in the *London Independent*, research by the British Royal Society for the

Protection of Birds (RSPB) has determined that sea-bird colonies off the Yorkshire coast and the Shetlands during 2003 "suffered their worst breeding season since records began, with many simply abandoning nesting sites" (2003, 12). The sea-bird populations are decreasing because sand eels are in decline. The sand eels feed on plankton, which have diminished as water temperatures have risen. This survey concentrated on kittiwakes, one breed of sea birds, but other species that feed on the eels, including puffins and razorbills, have also been seriously affected. Euan Dunn of the RSPB commented: "We know that sand eel populations fluctuate and you do get bad years. But there is a suggestion that we are getting a series of bad years, and that suggests something more sinister is happening" (p. 12).

Sand eels also comprise a third to a half of the North Sea catch, by weight. They have heretofore been caught in huge quantities by Danish factory ships, which turn them into food pellets for pigs and fish. During the summer of 2003, the Danish fleet caught only 300,000 English tons of its 950,000-ton quota, a record low (Sadler and Lean 2003, 12). The situation is "unprecedented in terms of its scale and the number of species it's affecting," said ornithologist Eric Meek of RSPB (Kaiser 2004, 1090).

DECLINES OF COD, SALMON, AND TROUT

Ocean waters generally warmed by an average of 0.06 degrees C between 1960 and 2000. Surface water, where many fish observed by humans live, has warmed more quickly, by as much as 0.31 degrees C. The North Atlantic in particular has been warming more quickly than any other oceanic region, with a temperature increase of 0.5 degrees C during the past twenty years, as warming accelerates (Connor [August 5] 2002, 12). Some deeper waters have not escaped temperature rises. Stebbing pointed out that the shelf-edge current, which flows from Spain to the British Shetland Islands at a depth of between 200 and 600 meters (656–2,000 feet), warmed by 2 degrees C between 1972 and 1992. He pointed out, according to a report in the *London Independent,* that, because this current runs north at a speed of about thirty-five miles a day, it is the most likely route for some tropical and semi-tropical species that have been observed in British waters (p. 12).

"As the world warms, the only way for wildlife species to live in the temperature they prefer is to move their ranges slowly poleward," said Stebbing, the lead author of the study published in the *Journal of the Marine Biological Association* (Connor [August 5] 2002, 12). "Fish are good indicators of temperature change because they are unable to regulate their temperature independently of the surrounding water. They therefore swim to keep themselves in waters of their preferred temperature range. Not only are changes in fish distribution likely to reflect temperature increases, but the arrival of new fish species are well monitored by fishermen, as well as scientists," Stebbing reported (p. 12).

During the 1930s, British fishermen harvested roughly 300,000 tons of cod (the staple of traditional English fish and chips) annually; by 1999, the catch was down to 80,000 tons ("Fished" 2000, 18). Spawning cod in the North Sea fell from about 277,000 English tons during the early 1970s to 54,700 English tons in 2001. Cod stocks in the North Sea have been in nearly continuous decline since the early 1970s; they fell below safe biological limits after 1984. Global warming and predation (overfishing) by humans and seals are major factors in the decline of the cod fishery. Another problem advanced by fishermen is dredging and the laying of pipes (Smith 2001, 7).

Research by Gregory Beaugrand and colleagues supports the idea that cod are declining in the North Sea not only because of overfishing but also because larval cod feed on plankton, which have diminished due at least in part to rising temperatures. They have written that "variability in temperature affects larval cod survival; [we conclude] that rising temperature since the mid-1980s has modified the plankton ecosystem in a way that reduces the survival of young cod" (Beaugrand et al. 2003, 661). Such observations have been supported by North Sea fishermen, who have been dragging rolls of silk behind their boats for seventy years to monitor the density of plankton populations. Given the plankton's decline, even a complete ban on cod fishing is unlikely to restore the fishery. Cod have declined not only in population but also in size. The peak of plankton abundance now occurs later in the year, after cod larvae experience their greatest need for them. This mismatch means that fewer larval cod develop into adults.

Rising water temperatures may drive trout and salmon from many U.S. waterways, according to a report from two environmental groups (www.defenders.org/publications/fishreport.pdf). Their study of eight species of salmon and trout in 2002 suggested that the cold-water habitat required by these species could shrink by more than 40 percent during the twenty-first century, given "business as usual" emissions of greenhouse gases. Salmon and trout, both cold-water species, are very sensitive to the temperature of their aquatic habitats. In many areas, these fish are already living at the margin of their temperature tolerances, meaning even modest warming could render a stream uninhabitable. Habitats for some of these species could shrink as much as 17 percent by 2030, 34 percent by 2060, and 42 percent by 2090, according to a study compiled by Defenders of Wildlife and the Natural Resources Defense Council ("Warming Streams" 2002). The groups' analysis covers four species of trout (brook, cutthroat, rainbow, and brown) and four species of salmon (pink, coho, Chinook, and chum). Researchers looked at air and water temperature data from more than 2,000 sites across the United States.

Anticipated increases in water temperatures vary by location, averaging 0.7–1.4 degrees F by 2030, 1.3–3.2 degrees F by 2060, and 2.2–4.9 degrees F by 2090, depending on future levels of greenhouse gases in the atmosphere. In addition to warming waters, wild trout and salmon populations are also under pressure from habitat loss due to human infrastructure development, competition with hatchery fish, invasive exotic species, and other reasons. "Now we must add climate change to the list of challenges they face," said Mark Shaffer, senior vice president for programs at Defenders of Wildlife. "If we don't address the cumulative impact of all these factors, we will see more of these populations switching from a recreational resource to being listed as threatened or endangered" ("Warming Streams" 2002).

A World Wildlife Fund (WWF) report issued during 2001 asserted that stocks of wild Atlantic salmon have been cut in half in two decades due to global warming and infections spread by hatchery-bred fish. The WWF report said that wild salmon already had disappeared from 309 of their 2,000 usual breeding areas around the world. Elizabeth Leighton, senior policy officer for the WWF in Great Britain, said: "When a river loses its salmon stock, that population is gone forever. The miracle of [salmon] returning to its spawning ground cannot be repeated" (Milmo and Nash 2001, 11).

The WWF report said that, as of 2001, 90 percent of healthy wild Atlantic salmon populations returning to Europe could be found in only three countries: Norway, Iceland, and the Irish Republic. Salmon have nearly disappeared from Germany, Switzerland, the Netherlands, and Belgium. Wild salmon are nearly extinct in Estonia, Portugal, Poland, the United States, and the warmer parts of Canada. Wild Atlantic salmon had nearly vanished by 2002 from many rivers in Scotland. Wild stocks were extinct in the rivers Shieldaig, Garvan, Attadale, and Sguord on the west coast of Scotland, with only a handful left in another ten rivers. The genetic integrity of the few remaining wild fish in Scotland's rivers also had been damaged through interbreeding between farmed and wild fish. A wide range of factors was thought to be affecting salmon stocks, including overfishing at sea and global warming. James Butler, director of the Spey Fishery Board, said: "In 14 rivers we found that stocks ranged from no fish to just a handful" (Cramb 2002, 7).

A marine parasite that began spreading among Yukon River Chinook salmon in 2002 can be traced in part to warmer stream waters that allow it to thrive. The parasite has ruined many of the fish, making them inedible, with a fruity odor and ruined meat. The illness, caused by a common micro-organism that targets ocean fish, was detected in about 35 percent of king salmon sampled in 2002 and 2003, said Richard Kocan, a fish pathologist overseeing a study for the federal Office of Subsistence Management. That's a significant increase over the number of infected fish found in 1999, 2000, and 2001. Kocan said that some Yukon Chinook now spend June and July migrating upstream through water warmed to 59 degrees or higher, temperatures that allow the parasite to spread faster and kill its hosts more quickly. "Some of these fish are swimming within a few degrees of lethal temperatures for healthy salmon, and we know many of them are infected," he said (O'Harra 2004, A-1).

The parasite was first noticed in a few salmon during the mid-1980s but grew much worse during the late 1990s. The culprit is a common protozoan called *Ichthyophonus*, which probably enters the fish through its food supply, then spreads into organs and flesh. The pathogen also has been detected in salmon in the Kuskokwim and Taku rivers, but it's not clear that it's precisely the same species found in the Yukon Chinook, according to Kocan's report. It's also been found in a few Yukon burbot, raising the possibility that it's now inside the freshwater system. The fish can be eaten safely; they are not toxic. "They taste bad, they smell bad and they look bad, but if you're starving, go ahead and eat them," Kocan said (O'Harra 2004, A-1). The parasite inhibits breeding, however, killing salmon before they spawn. An estimated 60 percent of the fish sick with the disease don't make it to spawning grounds.

LOBSTER CATCHES DECLINE IN WARMER WATER

During 2002, record high temperatures in New England were reflected in lighter-than-usual lobster catches offshore. Marine scientists attributed warmer-than-usual waters to seasonal aberrations, including a mild winter, a short spring, and a hot summer. Bob Glenn, a biologist at the state Division of Marine Fisheries, said he took a temperature reading of 77 degrees F in twenty feet of water off the coast of Bourne, Massachusetts, during the third week of August. The average reading is near 70 degrees F. "I'm not ready to jump on the global warming bandwagon, but this has been an exceptionally warm summer in many places off the Cape," Glenn said. "And this kind of water has a real impact on just about everything" (Healy 2002, B-1).

An account in the *Boston Globe* said:

The movements of lobsters, blue-fin tuna, and cod have become difficult to predict this summer [2002] because they are moving out of warmer water, several fishermen said. Even on the ocean side of the Cape, which typically has colder waters than its bay, lobsters haven't been appearing in some of the usual streams this month, and some lobstermen's businesses have taken a beating as a result. "I've lost $8,000 in two weeks because my traps aren't seeing nearly as many lobsters," said Billy Souza, who has 700 traps from Provincetown to Wellfleet and sells his catch out of his garage here. (Healy 2002, B-1)

Leaning against his truck after delivering about 150 pounds of lobster to his wife and chief saleswoman, Cheryl, Souza declared the day his best in weeks. His haul fell as low as twenty-five or thirty pounds on some days, when he mostly netted "eggers"—lobsters bearing eggs that must be thrown back—and he found himself baffled by the lobsters' movements. "I've been lobstering for 20 years, and from my own system these lobsters like it in 42-degree water," Souza said. "More and more of my traps are coming up totally empty" (Healy 2002, B-1).

Massachusetts state biologists associated the warm-water temperatures during the summer of 2002 to temperatures in Cape Cod Bay and Buzzards Bay the previous December, which were the warmest in twelve years, according to the state Division of Marine Fisheries. In one reading in sixty-five feet of water off Plymouth, the mean temperature was 46 degrees F in December, compared to the usual mean of 42 degrees F. In seventy-five feet of water off Buzzards Bay, December's monthly mean was 50 degrees F, compared to the normal reading of 45 degrees F (Healy, 2002, B-1). Temperatures in New England returned to average or below during the winters of 2002–2003 and 2003–2004 as pervasive cold and snow returned.

TROPICAL FISH AND WARM-CLIMATE BIRDS IN BRITISH WATERS

As cold-water fish abandon waters near Britain, species that usually live in southern waters have taken their places. Sightings of warm-water fish have been plentiful since about 1990 in British coastal waters and have received considerable publicity in local newspapers. Warm-water species have been turning up regularly off the coasts of Devon and Cornwall for more than a decade; some scientists believe they are clear indicators of global warming. In the late 1980s, southern species such as sunfish and torpedo rays began to appear; by the late 1990s, such visitors were no longer regarded as oddities. A series of small, cold-water marine animals such as copepods were replaced by their warm-water cousins. "As fish are very dependent on the temperature of the water, it is sensible to link these changes with changes in water temperature. They would be consistent with predictions of climate change," said Douglas Herdson of the National Marine Aquarium (McCarthy 2002, 13).

Such sightings are not unprecedented. What has changed is their number and frequency. With migrations of sea creatures have come a number of birds that feed on them. Bob Swann, secretary of the Seabird Group, an international conservation and research organization, said: "There is much evidence of species previously more associated with more southerly regions of the Atlantic appearing around the British Isles. Climate change can influence oceanic currents and the availability of food—a prime reason for the presence of these birds. Certainly a lot of our breeding seabirds are currently doing very well because they are finding plenty to eat" (Unwin 2001, 7).

A team of British marine biologists analyzed records dating back forty years and, in 2002, announced a "strong link" between the northward migration of fish and rising sea temperatures. These scientists related the arrival of tropical and semi-tropical fish off the coast of Cornwall, the southernmost tip of the British mainland, to increases in the average temperature of the North Atlantic Ocean. Surveying records to 1960, the scientists found that "more exotic species of fish are being caught or washed ashore now than ever before and that the sightings can be directly linked to a corresponding rise in sea temperatures." The link is said to be a "significant correlation" and could explain why Cornwall in particular has seen so many exotic species of marine wildlife from warmer regions of the world, according to Tony Stebbing, a biologist formerly with the Plymouth Marine Laboratory, whose work was funded by the Natural Environment Research Council (Connor [August 5] 2002, 12).

During May 2004, mantis shrimp (warm-water creatures usually found in tropical waters) were caught in trawler nets in Weymouth Bay, Dorset, on England's south coast. Fishermen took the orange-colored shrimp to the nearby SeaLife Centre, where experts identified them. The mantis (also known as "toe splitter") shrimp average three inches long. They can strike at up to 100 miles per hour with hammer-like claws, packing force as powerful as a .22-caliber bullet. Two years earlier, another colony of mantis shrimps was found at the north end of Cardigan Bay, Wales.

During midsummer 2004, a shoal of two-foot-long grey triggerfish was observed off the British coast. The triggerfish, whose usual habitat is the tropical Atlantic and the Mediterranean, were discovered two miles off the Isle of Purbeck, Dorset. During June 2002, a group of bright purple jellyfish, called "by-the-wind sailor fish," were found washed up at Kimmeridge Bay. Their usual habitat is the deep waters of the Mediterranean. Another visitor from southern seas was the four-inch weaver fish, which buries itself in sand close to shore and releases stinging venom from its dorsal spines when stepped on (Savill 2004, 7).

According to an account in the *London Independent*: "This summer's biggest seabird sensation was a red-billed tropic bird (*Phaethon aethereus*), which flew around a yacht about 20 miles south of the Isles of Scilly." This bird had not previously been recorded in northern European waters. The nearest colonies are on the Cape Verde Islands and islets off West Africa. Breeding also occurs on Ascension Island, St. Helena, the West Indies, and on the Red Sea, Persian Gulf, and Arabian Sea islands. Cape Verde and other islands off Africa are the source of the series of summer sightings of rare Fea's petrels (*Pterodroma feae*) off Scilly and the coasts of Devon and Cumbria (Unwin 2001, 7).

Whales and dolphins are being observed frequently in British waters as the area has warmed. During one ferry's round trip between Portsmouth and Bilbao in 2001, wildlife enthusiasts logged sightings of 71 fin whales, 4 Cuvier's beaked whales, 37 pilot whales, 25 common dolphins, 352 striped dolphins, 120 unidentified dolphins, and 7 unidentified large whales. In addition, two sperm whales were spotted in the English Channel from a ferry sailing between Portsmouth and

Bilbao (in northern Spain). Newquay's Blue Reef Aquarium took custody of two loggerhead turtles that probably drifted to Britain after damaging their flippers in the open ocean (Unwin 2001, 7).

Stella Turk of the Cornwall Wildlife Trust was quoted in the *London Independent* as saying that a rise in sea temperatures might be a cause of northward migration of tropical sea creatures and birds. "It appears the sea is becoming generally warmer and it could be that such creatures are staying close to Britain throughout the year. An unusually wide range has already been reported this summer and it's still only mid-July— there could be many more surprises over the coming weeks" (Unwin 2001, 7).

In addition to fish and birds, British observers in 2002 sighted hundreds of root-mouthed jellyfish, as well as rhizostoma octopus. Large, disc-shaped sunfish (*Mola mola*) also have been spotted more frequently in recent summers. Turk said that sightings in 2002 began earlier than before—in May. Another surprise is a report of a six-foot bluefin tuna, *Thunnus thynnus*. On July 14, 2002, about forty basking sharks were spotted a mile off Perranporth, North Cornwall. A pilot whale was observed at the same time. A spokesman for the Falmouth Coast Guard said: "The sharks are huge—they go up to 30 foot [long]. When the water warms up we do get basking sharks here, but it is unusual to get so many" (Unwin 2001, 7). The flying gurnard, unknown in British waters before 1980, also has been sighted more frequently. The gurnards, with their elongated pectoral fins that enable them to move quickly through the water, were first caught in the nets of Cornish fishermen. A sharp-nosed shark was first netted in 1984 (Connor [August 5] 2002, 12).

Global warming has brought an unexpected benefit to British West Country fishermen who have been struggling to make a living as their traditional catches dwindle. A slight rise in sea temperature has meant that valuable shellfish that once were unable to thrive north of the Channel Islands can now be farmed for export. Disc-like *Molluscs haliotis* (sea ear), which grow twenty centimeters (seven inches) long, have fetched high prices on Japanese markets. This valuable delicacy is known as abalone in most of the world but as "ormers" in Britain. Abalone is best known by tourists as a source of iridescent mother-of-pearl jewelry that is used as an inlay. In Japan, the gonads of the abalone eaten raw are regarded as a particular delicacy, while in California the abalone is consumed in steaks. The people of Guernsey, in England, have traditionally eaten ormer stew (Brown and Sutton 2002, 8).

One notable warm-water visitor to English waters, a single slipper lobster, was caught near the southwest tip of England and later displayed at the Plymouth National Marine Aquarium. The five-inch-long slipper lobster (*Scyllarus arctus*) is usually found near the coasts of the Mediterranean Sea. Only about a dozen have been recorded in United Kingdom waters during the last 250 years. The most recent specimen, caught by fisherman Barry Bennett, was the fifth to be caught in British waters since 1999. The increasing number of slipper lobsters in British waters is one of many indications that warm-water marine species have been moving northward because ocean temperatures are rising (McCarthy 2002, 13).

Other warm-water fish have been found in English waters from two fish families, the breams and the jacks. A Guinea amberjack was first recorded off Guernsey. The previous year there were five sightings of the almoco jack in the West Country (McCarthy 2002, 13). A tropical zebra sea bream, usually resident off the West African coast, was caught, for the first time, during 2002 in British waters. The fourteen-inch fish was accidentally netted by Ross White (aged twenty-nine), a commercial fisherman, near Portland, Dorset, far from its native waters off Senegal and Mauritania. A seahorse also was sighted in the Thames River—a rare, but not unprecedented, event. Another seahorse was sighted in 1976.

During November 2001, Britain's first reported barracuda was caught about forty miles from the site of the slipper lobster catch. England is not alone in observing tropical fish; two years after the English barracuda catch, another barracuda was caught on Seattle's waterfront. As they have near England, cold-water cod (of the Pacific variety) are becoming increasingly rare off the state of Washington's coastline (Stiffler and McClure 2003, A-8). During the summer of 2004, a giant squid, a species that usually ranges no further north than Mexico's coastal waters, was caught near Maple Bay in southern British Columbia ("Jumbo Squid" 2004).

Charles Clover of the *London Daily Telegraph* reported that red mullet, which were restricted to waters south of the English Channel before 1990, by 2002 were being caught in commercial quantities on both coasts of Scotland. "The largest geographical movement recorded over the past decade, the warmest on record," wrote Clover, "has been made by species of zooplankton, tiny shrimp-like creatures which form the base of the marine food chain. Warm-water species of copepod, as these crustaceans are known, have moved 600 miles northwards up the Bay of Biscay over the past decade, bringing warm-water fish species with them" (2002, 14). At the same time, the cold-water copepod, *Calanus finmarchicus*, the main food of the cod and of the sand eel on which the cod also feed, has moved north from the North Sea. Martin Edwards, of the Sir Alastair Hardy Laboratory in Plymouth, said the North Sea was in a "transitional state," with the consequences for fish stocks, already endangered as a result of overfishing, hard to predict (p. 14).

WARM-WATER SPECIES OFF IRELAND

Warm-water fish have been sighted in Irish waters as well. Among these have been various species of sharks, poisonous puffer fish, loggerhead turtles, and triggerfish. According to a report by Lynne Kelleher in London's *Sunday Mirror*, coastal waters that previously reached a summer maximum of 15 degrees C now commonly reach 20 degrees C, drawing the warm-water species. According to Kelleher, "Fish never seen before are appearing in greater numbers and some are beginning to breed as they become acclimatized. A great white shark was spotted off the coast of Cornwall in recent years and exotic fish such as moray eels, mako sharks and anchovies are swimming off Irish shores" (Kelleher 2002, 15).

Kevin Flannery, a marine biologist with the Department of Marine and Natural Resources, said the fish are arriving because of warmer water temperatures. He commented:

> There is a definite correlation between the rise in temperatures and the arrival of new species of fish in Irish waters. . . . The temperatures used to go below five degrees in the winter and go up to 14 or 15 degrees in the summer. Now they are between five and seven in the winter and can go between 17 and 20 degrees in the summer. A moray eel, which is found in Australia and Canada, was caught by a vessel from Waterford off the coast of Cork last year. We have found three puffer fish. One was found off Fenit in Kerry last winter. A number of loggerhead turtles have been found washed up thousands of miles from where they came. (Kelleher 2002, 15)

Large numbers of anchovies, usually found off the coast of Portugal, have been observed swimming in waters from Shannon to Kinsale, in Irish waters. Puffer fish, which are poisonous, also have been sighted, probably for the first time. Tropical triggerfish, which have been found in the largest numbers, are small in size but become aggressive to divers and swimmers if they are disturbed during the breeding season. Different types of bream and dory fish also have been sighted in Irish waters. Flannery said:

> The number of tropical species has increased dramatically. There are between 10 to 15 rare species found in recent years. One [fishing] vessel found about 80 triggerfish last year. They can do a lot of damage here to crab, lobster and crayfish that they feed on. They have teeth like a rat and can kill a lobster. Fishermen are finding them inside the lobster pots. Their natural habitat is the tropical waters of Spain and Africa. They won't survive in waters less than 14 degrees. In 1995, the waters reached temperatures of about 20 degrees that was one of the highest temperatures. The triggerfish are staying around. They are not just coming in and out

to feed. We are finding a number of pregnant triggerfish. This suggests they are acclimatizing. They wouldn't breed unless they were staying around. (Kelleher 2002, 15)

Flannery said that, as tropical fish migrate to Ireland, traditional cold-water species such as cod have been leaving its coastal waters. He said: "The rise in temperature could also mean the demise of cod which need cold temperatures. If temperatures go over 17 degrees they will die. There has also been the demise of the Arctic char in our lakes. It is a combination of the pollution and the temperature" (Kelleher 2002, 15).

TROPICAL FISH AND SHARKS OFF MAINE

A few sharks were sighted off the coast of Maine during the summer of 2002, as they followed warmer waters. According to an account in the *Boston Herald*:

Whether it was blue or mako sharks that invaded the crowded Maine beach, banishing swimmers to the sand, the unusual occurrence is the result of either fish chasing food close to shore or warmer waters luring a more tropical variety to New England waters, marine experts say. The Wells, Maine, shark sightings are "a very unusual circumstance," said Greg Skomol, a shark specialist for the state Division of Marine Fisheries. "We wouldn't expect to see that in Maine or in New England, in general." (Richardson 2002, 3)

The sharks swam near the beach three days in a row at a popular resort called Vacationland as the tide was receding, probably hoping to catch whatever baitfish were being washed out of nearby crevices in a rocky ledge. The sharks' usual food supply (mackerel and some herring) may have been lured to shallow waters by unusually warm air temperatures in the nineties. "We had hot weather and sometimes that will change the distribution of fish and bait fish that sharks feed on," Skomol said. "The high temperatures may have increased the mackerel and herring, therefore bringing in the sharks. But with one event, it's hard to predict anything, though it's worth watching" (Richardson 2002, 3).

Bruce Joule, a marine biologist in Boothbay Harbor who advised officials to close the beach when sharks appeared, said that he believes water temperature had nothing to do with the sharks. "I don't think the temperature has really anything to do with it," Joule said. "The sharks came in following bait, whether it was natural migration [of bait fish] or the smaller fish were chasing their own food. It's nature" (Richardson 2002, 3). Following Joule's advice, Wells Fire Chief Marc Bellefeuille closed the beach as a precautionary measure, allowing sunbathers into the water only up to their ankles. "This is the first time anyone can remember shark sightings like this," Bellefeuille said (p. 3).

At about the same time, unusually warm water temperatures also brought tropical fish to New England shores, including yellow fin, dolphin fish, white marlin, and sometimes skipjack tuna. According to Skomol, "how close they come depends on the water. They're not aggressive. We've been seeing them the last three years" (Richardson 2002, 3). President George W. Bush was photographed during the summer of 2002 landing a large striped bass that his daughter Jenna had caught on a fishing holiday off the coast of Maine. Striped bass is a warmer-water species, which only a generation ago would never have been seen so far north along America's Atlantic Coast (Connor [August 5] 2002, 12). Bush displayed no clue that he realized his catch was unusual or influenced by climate change.

SHARKS IN ALASKAN WATERS

An increasing number of sharks in Alaskan waters, along with de-clining seal and sea lion populations, could be a result of global warm-ing. A spotter in an Alaska Fish and Game plane on a survey of sea otters observed hundreds of shark fins in the small bay near Port Gravina, Prince William Sound. More sharks massed unseen below the surface, feasting on salmon returning to spawn in nearby rivers and creeks. "That aerial count would be a high number of sharks in one spot for any place in the world," said Vince Gallucci, University of Washington

A salmon shark landed in Alaskan waters. Courtesy of Bruce Wright, Conservation Science Institute.

professor of fisheries and aquatic sciences. Gallucci, who has studied shark population dynamics for more than a decade, the last two years in Alaska, organized an American Association for the Advancement of Science session, "Not Enough Sea Lions, Too Many Sharks: Global Warming Signal?" The number of Pacific sleeper sharks encountered by halibut fishing vessels has increased every year since 1997, more than doubling, according to a database made available to Gallucci by the Pacific Halibut Commission. "Fishermen wouldn't forget 200 pound animals bending hooks and wrecking their nets," Gallucci said, displaying a number-three steel, halibut circle hook that had been almost straightened by a salmon shark ("Sharks" 2002).

Fieldwork by Gallucci and colleagues, along with wildlife surveys by others, led Gallucci to assert that top predators in Alaska's sub-Arctic waters have shifted to a new balance, with sharks outnumbering pinnipeds (that is, animals with finned feet such as sea lions and seals). "Increases in salmon sharks and Pacific sleeper sharks, both sub-arctic northeast Pacific shark species, don't represent ecological invasions and they aren't range extensions since both sharks are endemic," Gallucci argued ("Sharks" 2002). He believes that population changes are tied to decades-long swings in climate and continuing global warming, which have helped change the populations of fish that these animals eat. "Sharks, being the more efficient eaters, just may be able to take greater advantage of changes in the food that's available," Gallucci said (2002).

ANCHOVIES SPREADING NORTHWARD

The anchovy, usually found in the Mediterranean Sea, has been caught by several fishing vessels as far north as Donegal Bay in Ireland. Four vessels working on mackerel and herring during December 2001 reported catches of anchovies, with one of the most substantial hauls made in Donegal Bay. John Molloy, a scientist with Ireland's Marine Institute, told the *Irish Times* that one vessel working out of Greencastle, County, also caught some off Malin Head. "It is very unusual, particularly so far north," Molloy said. "Occasionally you'd get individual specimens in trawl hauls, but nothing of any note" (Siggins 2001, 1).

Kevin Flannery, a sea-fishery officer with the Department of the Marine, said that anchovies had also been caught off the Old Head of Kinsale. "It would seem that the shoals are staying well offshore and may be swimming within a warm movement of water that had been carried up north of Biscay and is holding this area of high pressure with it," Flannery said (Siggins 2001, 1).

CORAL REEFS "ON THE EDGE OF DISASTER"

The scope of coral devastation from climate change and other human impacts rivals the losses endured by the flora and fauna of the world's great rainforests. According to a number of estimates, half of the world's coral reefs may be lost by 2025 unless urgent action is taken to save them from the ravages of pollution, dynamite fishing, and warming waters. Many of the coral reefs that are falling prey to human-induced destruction are among the largest living structures on Earth. Many are more than 100 million years old. Coral that has lost its ability to sustain plant and animal life turns white, as if it had been doused in bleach. Afterward, the dead coral often is subsumed by a choking shroud of grey algae. Since coral polyps and their calcium carbonate skeletons "are the foundation of the entire ecosystem, fish, mollusks, and countless other species, unable to survive in this colorless graveyard, rapidly disappear, too" (Lynas 2004, 107).

Aside from the obvious ravages of fishermen who blast the reefs and pour cyanide on them, the reefs are also threatened by ocean temperature spikes that many marine biologists attribute to global warming and short-term climate phenomena such as El Niño. Most corals live very close to the upper limit of their temperature tolerances. Temperature increases of only a few degrees over a sustained period cause the death of symbiotic micro-organisms living within the coral tissue that provide energy for the colony (Hirsch 2002, 14).

Warnings of the world oceans' coral holocaust have been widespread in the scientific literature. The journal *Science*, for example, devoted a cover story to the subject on August 15, 2003, wherein T. P. Hughes and colleagues concluded that "the diversity, frequency, and scale of human impacts on coral reefs are increasing to the extent that reefs are threatened globally" (p. 929). Prominent among these impacts is anthropogenic warming of the atmosphere and oceans that may exceed limits under which corals have flourished for a half-million years. Some types of corals are more vulnerable to warming than others, however, so "reefs will change rather than disappear entirely" (p. 929). Rising ocean temperatures, however, will certainly reduce biological diversity among corals.

Clive Wilkinson, representing the Australian Institute of Marine Science, said during 2000 that 27 percent of reefs worldwide, crucial nurseries for fish and plants, had died or were in serious trouble following record sea temperatures two years earlier. Another 25 percent will die during the next twenty-five years if existing trends continue, he said. Wilkinson, the coordinator of the Global Coral Reef Monitoring Network, cited a slight cause for optimism, however, thanks to the new priority given to coral conservation by some governments. "Until recently, governments were in denial. That's no longer the case but we have a long way to go to reverse the trend," Wilkinson said. He noted, however, that "the climate-change models suggest things are going to get a lot worse for corals in most regions" (Nuttall [October 25] 2000).

Although individual governments might be able to curtail destructive fishing practices and pollution, the gradual, pervasive warming of the oceans that kills corals is another matter. Worldwide, coral bleaching accelerated in 2002 at its worst pace since 1998, paralleling a rise in global temperatures. The year 2002 became the second worst for bleaching after the major El Niño year of 1998. More than 430 cases of coral bleaching were documented that year, notably from the Great Barrier Reef in Australia as well as from reefs in countries including the Philippines, Indonesia, Malaysia, Japan, Palau, Maldives, Tanzania, Seychelles, Belize, Ecuador, and off the Florida coast of the United States ("New Wave" 2002). Reports from the World Fish Center and the International Coral Reef Action Network (ICRAN) all documented the increasing toll on the world's corals. Coral bleaching has become endemic around the world. A survey of reefs in the Seychelles in January 1999, for example, showed that 80 percent of the coral was dead, with more than 95 percent mortality in some areas (Leggett 2001, 324). (ReefBase, which can be viewed at www.reefbase.org, includes more than 3,800 records going back to 1963 that include information on the severity of bleaching.)

Callum M. Roberts and colleagues, writing in *Science*, sketched the scope of possible extinctions faced by the world's coral reefs:

Analyses of the geographic ranges of 3,235 species of reef fish, corals, snails, and lobsters revealed that between 7.2 per cent and 53.6 per cent of each taxon have highly restricted ranges, rendering them vulnerable to extinction.... The 10 richest centers of endemism cover 15.8 per cent of the world's coral reefs (0.012 per cent of the oceans) but include between 44.8 and 54.2 per cent of the restricted-range species. Many occur in regions where reefs are being severely affected by people, potentially leading to numerous extinctions. (2002, 1280)

Klaus Toepfer, executive director of the U.N. Environment Programme (UNEP), has noted: "Coral reefs are under threat worldwide from a variety of pressures, including unsustainable fishing methods, such as dynamite and cyanide fishing, insensitive tourism, pollution and climate change. Every effort is needed to conserve these vital habitats for fish and other marine life for the benefit of local people who rely upon them for protein and livelihoods" ("New Wave" 2002). Toepfer added, "Coral reefs may be the equivalent of the canaries in coal mines giving early warning that the world's ecosystems can no longer cope with growing human impact" (Lean 2001, 7).

A study involving more than 5,000 scientists and divers who monitored coral reefs adjacent to fifty-five countries for five years found only one reef out of more than 1,100 that was judged to be in near-pristine condition. "Coral reefs have suffered more damage over the last 20 years than they have in the last one thousand," said Gregor Hodgson, author of the report and a University of California at Los Angeles visiting professor who heads Reef Check, a monitoring program based at UCLA's Institute of the Environment. "It is the rate of decline and the global extent of the damage that is so alarming, with species reasonably abundant 30 years ago now on the verge of extinction" (Hirsch 2002, 14). Hodgson said that the Reef Check project is designed to be an early-warning system and that he hopes other scientists will examine the issues raised in his report. "Although it is a big study, we have looked at only a tiny fraction of the world's reefs," he commented (p. 14).

Many species of fish, shrimp, and lobster were missing from the majority of the reefs under study, as fishermen poured cyanide and blasted dynamite to harvest edible species from them. Explosives are the most destructive fishing method for reefs. As they explode on the surface of the water, shock waves kill a majority of fish species on the reef and cause severe damage to its structure. Nitrogen fertilizers from coastal communities are destroying many reefs as well, feeding large, fleshy algae that choke hard, stony corals that comprise the foundations of many reefs (Hirsch 2002, 14). Larger predatory fish, including the Nassau grouper, Barramundi cod, and hump-head wrasse, were missing from at least eight of every ten reefs surveyed. These big fish are prized in Hong Kong and other East Asian markets. Fishermen in Indonesia and the Philippines often stun these fish with cyanide, also killing other nearby animals as well as the coral.

Most of the coral reefs of the world's oceans will disappear within thirty to fifty years, according to Rupert Ormond, a marine biologist at Glasgow University marine biological station at Millport in Scotland. Ormond told the British Association science festival in Glasgow that global warming will raise ocean temperatures to levels that will bleach the great reefs of the Pacific and Indian oceans, the Caribbean, and the Red Sea.

> To begin with, we only saw these events in El Niño years, when the ocean temperatures tend to be warmest. . . . Within 10 to 20 years we will get massive bleaching on a wide scale almost every year. One can predict, looking at those figures, that maybe within 50 years there will be very little left of corals in coral reef countries. . . . Frankly, I find the whole prognosis extremely gloomy. I cannot see what can be done, given that there is something like a 50 year time lag between us trying to control carbon dioxide emissions and the temperature of the oceans beginning to drop. (Radford [September 6] 2001, 9)

Ormond said that rising ocean temperatures are leading to an irreversible decline of the great coral reef barriers of the tropics, "from the Indian Ocean to the Caribbean and covering the entire Pacific Ocean." He believes world coral reefs will be substantially dead within fifty years (Woodcock 2001, 4).

THE WORLDWIDE NATURE OF CORAL REEF DECAY

The threats to coral reefs are similar around the world. More than 80 percent of Indonesia's coral reefs, for example, have been threatened mainly due to blast-fishing practices and bleaching, according to the U.N. Environment Programme's (UNEP's) new *World Atlas of Coral Reefs* (Spalding 2001). Indonesia, along with the Philippines, Malaysia, and Papua New Guinea, with between 500 and 600 species of coral in each of these countries, is home to the world's most diverse range of corals ("Over 80" 2001).

"Our new atlas clearly shows that coral reefs are under assault," said Toepfer. He continued, "They are rapidly being degraded by human activities. They are over-fished, bombed and poisoned. They are smothered by sediment, and choked by algae growing on nutrient rich sewage and fertilizer run-off. They are damaged by irresponsible tourism and are being severely stressed by the warming of the world's oceans. Each of these pressures is bad enough in itself, but together, the cocktail is proving lethal" ("Over 80" 2001).

The newly published *World Atlas of Coral Reefs* provides detailed descriptions of the coral reefs in every country in the world. "And," wrote Mark Spalding, lead author of the *Atlas*, "in every country the same problems are present" (Spalding [September 12] 2001, 8). Traveling in the Indian Ocean, Spalding wrote in the *London Guardian*,

Over the next six weeks we watched the corals of the Seychelles die. Corals are to reefs what trees are to forests. They build the structure around which other communities exist. As the corals died they remained *in situ* and the reefs became, to us, graveyards. Fine algae grows over a dead coral within days, and so the reefs took on a brownish hue, cob-webbed. In fact, the fish still teemed and in many ways it still appeared to be business as usual, but as we traveled—over 1,500 kilometers across the Seychelles—the scale of this disaster began to sink in. Everywhere we went was the same, and virtually all the coral was dying or already dead.... What I witnessed in the Seychelles was repeated in the Maldives and the

Chagos Archipelago. In these Indian Ocean islands alone, 80 to 90 per cent of all the coral died. (p. 8)

Warmer sea-surface temperatures have killed 90 percent of the coral reefs near the surface of the Indian Ocean in only one year, while the remaining 10 percent could die in the next twenty years, according to work by Charles Sheppard of England's University of Warwick. During 1998, according to Sheppard, rapid warming devastated shallow-water corals from the surface to 130 feet below sea level. Some of these corals began to recover in subsequent years, but the risk continues, according to Sheppard. "It's like a forest," he said. "If you kill off 90 per cent, there might be just enough left to sustain some life around it, such as squirrels and so on. But if you have the same impact again and again there's no clear line as to when it's alive or not as a forest" (Arthur [September 18] 2003).

In research published by *Nature*, Sheppard used a computer model for global warming developed by Britain's Hadley Centre for Forecasting to anticipate how sea temperatures will affect coral survival (Sheppard 2003, 294–297). "Most corals don't mature until [they are] five years old and, five years since the 1998 event, most sites have recovered only marginally," Sheppard said (Arthur [September 18] 2003). According to Sheppard's analysis, a proportion of a given coral dies with each annual peak in the sea temperature. "The warming trend is only a fraction of a degree each year, which is swamped by the annual change," Dr. Sheppard said. "But south of the Equator the probability of the temperature rising enough to kill all the corals means they could be wiped out as soon as 2020" (Arthur [September 18] 2003).

Worldwide, according to Gian-Reto Walther and colleagues writing in *Nature*,

six periods of mass coral bleaching have occurred since 1979 and the incidence of mass coral bleaching is increasing in both frequency and intensity. The most severe period occurred in 1998, in which an estimated 16 per cent of the world's reef-building corals died. The impact of thermal stress on reefs can be dramatic, with the almost total removal of corals in some instances. . . . Climate change is apparently affecting the reproductive grounds of krill, and consequently its recruitment, by reducing the area of sea ice formed near the Antarctic Peninsula, which leads to both food web and human economic consequences. (2002, 393)

Walther and colleagues continued: "The evidence indicates that only 30 years of warmer temperatures at the end of the twentieth century have affected the . . . organisms, the range and distribution of species, and the composition and dynamics of communities. . . . The implications of such large-scale consistent responses to relatively low rates of climate change are large and the projected warming for the coming decades raises even more concern about its ecological and also socio-economic consequences" (p. 393).

THE DEMISE OF AUSTRALIA'S GREAT BARRIER REEF

Australia's 1,200-mile-long Great Barrier Reef is the largest coral reef system in the world, including more than 2,600 individual reefs and about 300 islands. The Great Barrier Reef, which has been described as the Amazon jungle of the marine world, has been severely degraded, however, having suffered during 2002 its worst bleaching event on record ("New Wave" 2002). The destruction of the Great Barrier Reef has been accelerated not only by warming water temperatures but also by agricultural runoff, including fertilizers, herbicides, and pesticides, which kills corals. Runoff also causes increases in algal blooms, which feed increasing populations of crown-of-thorns starfish that devour corals.

Extensive areas of Australia's Great Barrier Reef showed no signs of recovery a year after extensive bleaching during the summer of 2001–2002. At that time, aerial surveys showed that about 60 percent of the reef's 6,700 square kilometers had been affected. According to Ray Berkelmans, a research scientist at the Australian Institute of Marine Science in Townsville, recovery was "poor to non-existent. . . . What had been beautiful reef is now acres and acres of dead coral covered with algae" (G. Roberts 2003, 4). Berkelmans and other scientists expressed concern that another hot summer could damage the reef even more. The Great Barrier Reef has experienced six bleaching events since 1980.

By 2004, the Great Barrier Reef had lost about half its coral cover, compared to the 1960s. During the 1960s, about 40 percent of the reef was covered with corals; by 2004, the coverage averaged 20 percent (Williams 2004, 11). Australian scientists said that this decline was attributable entirely to human impact, most notably from rising water temperatures due to global warming, overfishing, and water pollution. David Bellwood of James Cook University in Townsville, coauthor of a major review published in *Nature* (Bellwood et al. 2004, 827–833), said that the loss of coral was not a surprise to scientists who knew of the damage done by three major outbreaks of crown-of-thorns starfish since the 1960s and two large-scale bleaching events in 1998 and 2000. "Data has been accumulating for years on this and we've now gotten around to pulling it all together and looking at the overall pattern," he said (Williams 2004, 11).

A report by the Environment News Service stated: "Coral bleaching in the Great Barrier Reef Marine Park may be the worst on record, scientists said in late May, 2002, after the most comprehensive aerial survey ever conducted. The survey is aimed at helping unravel the implications of global warming for reef management" ("Pacific Too Hot" 2002). "Our aerial surveys found that nearly 60 percent of the reef area in the marine park was heat stressed to some extent as indicated by bleaching," said Berkelmans. "Until now, the coral bleaching episode in 1998 was the worst on record, but the 2002 event was probably worse because more reef area was affected" (2002).

The survey by scientists from the Australian Institute of Marine Science, CRC Reef, and the Great Barrier Reef Marine Park Authority examined more than 640 locations from the northern tip to the southern end of the Great Barrier Reef Marine Park using light aircraft. The team also used SCUBA to confirm results and determine whether corals were likely to recover from bleaching or would die. The aerial surveys indicated that bleaching was worst in the Princess Charlotte Bay region, near the Turtle Island Group, on inshore reefs from Cape Upstart to the Whitsundays, in some reefs in the Sir James Smith Group, and in the Keppel Island area. Moderate to very high bleaching was observed inshore and offshore from around Cape Flattery to Mackay. "Our underwater surveys found that few reefs escaped bleaching, but it appears likely that most reefs will recover with only minor death of corals," said Paul Marshall of the Great Barrier Reef Marine Park Authority, who led the underwater surveys. "We did find that some of the most severely bleached reefs were devastated with 50 percent and 90 percent of coral dead at some sites" ("Pacific Too Hot" 2002).

Several toxic pollutants carried from Australian farms by floods have been threatening inshore areas of the Great Barrier Reef, according to a study by marine experts. The report, by the Great Barrier Reef Marine Park Authority, found that pollution levels in floodwaters, or flood plumes, were four times worse than fifteen years previously (Freeman and Cowie 2002, 7). The newspaper account of this study reported, " 'Of about 750 inshore reefs in the park designated as a world heritage site by the United Nations, 200 were considered at high risk and more than 400 were at risk,' said Sheriden Morris, the park authority's water-quality director" (p. 7). This study said that floodwaters occasionally surge onto the reefs from twenty-six river systems in Queensland, carrying sediment, nutrients, herbicides, and pesticides. Sediment and nutrients damage sea-grass beds and create algae growth, while pesticides and herbicides stop the growth of plankton and sea grass, the report said. The study found that concentrations of dissolved nutrients in flood plumes were well above levels known to damage coral reef ecosystems. Imogen Zethoven, World Wildlife Fund Great Barrier Reef campaign manager, said that most of the pollution comes from agricultural land. "You're getting high levels of fertilizer run-off from sugar cane growing properties and horticultural properties.... We've got to find a way of growing cattle, cane and fruit and vegetables where those industries can co-exist harmoniously with the tourism and fishing industries," she said (p. 7).

By the summer of 2002, authorities in Australia were issuing individual heat-wave alerts for events that they believed could harm the

Great Barrier Reef. Pervasive warming of shallow waters above the reefs was the most immediate concern. In the meantime, Australia's federal government, following the example of the United States under the administration of George W. Bush, was refusing to ratify the Kyoto Protocol, even as heat records were broken, wildfires nipped at the capital of Canberra, and the magnificent Great Barrier Reef decayed.

Australian Institute of Marine Science senior principal researcher John Veron said during 2002 that the Great Barrier Reef would be so severely degraded in fifty years by global warming and other anthropomorphic insults that "my grandchildren won't see a Great Barrier Reef like I did, that's for sure.... It will be mostly dead." He continued, "Words don't exist to describe what's just around the corner—coral reefs, more than anything else, are first in line for the effects of global warming. It's too late for a lot of the areas like the Great Barrier Reef and the reefs off Western Australia; I can't see any escape from that conclusion" ("Warming Doom" 2002). Coral bleaching, which was considered dangerously severe at the beginning of the twenty-first century, may be regarded as mild within fifty years, Veron warned (2002).

In contrast to Veron's bleak assessment, a few other reports assert that the Great Barrier Reef is improving due to better management. These reports are a distinct minority, however. One such report trumpeted a belief that "fears for the future of Australia's Great Barrier Reef were ... laid to rest with the revelation that it is now one of the world's healthiest coral reefs" ("Great Barrier" 2002, 11). According to the Australian Institute of Marine Science, this newspaper reported, "Only about six per cent of the vast reef ... is now suffering from the phenomenon. The government's Great Barrier Marine Park Authority is credited with the restoration. It worked tirelessly to ensure water quality, protect fish stocks and set up marine sanctuaries" (p. 11). The article quoted Wilkinson as saying, "Reefs, if they are left alone and not stressed, will recover quite rapidly. Our first state-of-the-reef report in 1998 identified mass coral bleaching, which killed off about 16 per cent of the world's coral stocks. The latest report shows there has been recovery but it is in areas that are quarantined from other activity" (p. 11).

In addition to the destruction of the Great Barrier Reef, warming oceans are playing a major role in the decline of another Australian marine landmark, the giant kelp forests of Tasmania. Two-thirds of the kelp beds along Tasmania's east coast have died during the last fifty years, probably as a result of a 1.5–2.0 degree C rise in water temperatures. Kelp generally dies if waters rise above 20 degrees C. Additionally, warm-water species that feed on kelp (such as sea urchins) have been moving in, reducing, according to marine ecologist Craig Johnson of the University of Tasmania, "rich, luxurious, highly diverse seaweed beds to barren, over-grazed 'moonscapes.' " The kelp forests are a major tourist attraction and habitat for rock lobster and abalone ("Kelp Points" 2004, A-8).

THE CONDITION OF CORALS NEAR FIJI

David Bellamy, a British botanist, has warned that many of Fiji's spectacular coral reefs are being ruined as a result of bleaching caused at least partially by global warming. Bellamy visited Fiji during 2001 to inaugurate a project by a British charity to survey reefs in the Mamanuca Islands off the west coast of the main island of Viti Levu. Coral Cay Conservation was invited by members of the tourism industry in Fiji to undertake a three-month pilot project. "Hoteliers contacted us saying they wanted to find out what had gone wrong, because without reefs there's nothing to attract high-spending divers," said Bellamy (Miles 2001, 4). In addition to bleaching, reefs near thirteen Mamanuca resorts have been threatened by overfishing, which results in an imbalance in the ecosystem, Bellamy said. Andrea Dehm, manager of Ovalau Watersports on Viti Levu, said the bleaching had caused diving in the area to become increasingly disappointing. "Our divers used to comment on the beautiful colors, but now it's all white," she said. "People don't want to see that and some operators are closing" (p. 4).

CORAL CATASTROPHE IN THE CARIBBEAN

Four-fifths of the coral on Caribbean reefs have disappeared during twenty-five years. Human provocations, including but not limited to warming seas, are responsible for most of the destruction. The scope of the loss has astonished even scientists who have been studying the global decline of coral. The destruction has been unmatched for several thousand years, according to a study in the journal *Science* (Gardner et al. 2003; McCarthy 2003, 3). The team leader of the study, Isabelle Côté, a French-Canadian specialist in tropical marine ecology, said that the causes of the corals' decline include industrial, agricultural, and other human pollution, overfishing, diseases, stronger storms, and higher sea temperatures (McCarthy 2003, 3).

The work was carried out by researchers at the University of East Anglia, United Kingdom, and its associated Tyndall Centre for Climate Change Research, using data from 263 Caribbean sites, including Mexico, Barbados, Cuba, Panama, the Florida Keys, and Venezuela. "We report a massive region-wide decline of corals across the entire Caribbean basin," the scientists wrote in *Science* (Gardner et al., 2003). The reefs of the Caribbean, seriously weakened by human depredation, may now be unable to withstand future warming. "The ability of Caribbean coral reefs to cope with future local and global environmental change may be irretrievably compromised," the team reported (McCarthy 2003, 3).

The study involved hard corals, the tiny animals that slowly build coral reefs from the calcium carbonate that they excrete. The scientists found that in 1977, the start of the survey period, 50 percent of a typical Caribbean reef was covered in live corals, which is considered healthy. By 2002, however, a typical reef was 10 percent covered, which is regarded as potentially fatal. "The end result surprised us, as well as all the people who gave us data," said Côté. "The rate of decline we found exceeds by far the well-publicized rates of loss for tropical forests" (Gardner et al., 2003; McCarthy 2003, 3).

To provide one of many examples, *London Independent* environmental writer Geoffrey Lean described the fate of corals off one stretch of Jamaica's coastline:

Discovery Bay in Jamaica was once one of the finest in the Caribbean but now its beauty and rich wildlife have been smothered by algae as a result of a complex mixture of man-made and natural catastrophes. These began [with] over-fishing...soon after the island's colonization.... Over-fishing [undermined the health of the reef],...laying it open to a series of catastrophes in the 1980s. The first affected a species of sea urchin which had grazed on the algae and kept it in check after the fish were depleted. It was hit by a mystery disease in 1983–4 (brought, according to one theory, in ballast water in a ship passing through the Pacific Canal) and began dying out. Then the coral itself was hit by diseases and a hurricane did even greater damage. Now, scientists say, it can hardly be called a reef at all. (Lean 2001, 7)

Among Caribbean corals, according to Toby A. Gardner and colleagues, "average hard cover [has been] reduced by 80 per cent, from about 50 per cent to 10 per cent cover, in three decades." The ability of Caribbean coral reefs to cope with future local and global environmental change, they wrote, "may be irretrievably compromised" (2003, 958).

Corals may recover if conditions change. This study, while so pessimistic about the prospects of the corals as a whole, found that some areas of degraded coral in the Caribbean appear to be recovering.

The bad news, however, is that the new coral communities seem to be different from the old ones.... At this point, we do not know how well these new assemblages will be able to face new challenges, such as rising sea levels and temperatures as a result of global warming. .... Given current predictions of increased human activity in the Caribbean, the growing threat of climate change on coral mortality and reef framework-building, and the potential synergy between these threats, the situation for Caribbean coral reefs does not look likely to improve in either the short or the long term. (McCarthy 2003, 3)

CORAL REEFS AND "WHITE POX"

An epidemic of "white pox" that has nearly destroyed the once-plentiful elkhorn coral in some areas of the Caribbean Sea has been traced to bacteria found in sewage, scientists reported in the June 25, 2002, edition of the *Proceedings of the National Academy of Sciences* (Patterson et al. 2002, 8725–8730). The epidemic has reduced populations of elkhorn coral by as much as 70 percent in waters surrounding some of the Florida Keys. "It is very sad that the one coral species affected is the magnificent branching elkhorn coral," said University of Georgia ecologist James Porter. "These are the giant redwoods of the reef," he added. "What used to be the most common coral in the Caribbean has now been recommended for inclusion on the endangered species list" (Nesmith 2002, 3-A).

Porter and colleagues have established that the epidemic is caused by bacteria called *Serratia marcescens*, found in the digestive tracts of humans and many other animals. The disease, which is extremely contagious, causes pale blotches on coral that is usually tan-colored. Katherine Patterson, a researcher who works with Porter, said, "Identification of a fecal enteric bacterium as the cause of white pox means we cannot blame global warming as the main problem on the coral reefs, but it all adds up. Warmer water depresses coral growth, but increases bacterial growth. In combination, this domino effect could foretell a disaster" (Nesmith 2002, 3-A). Warmer waters may thus be turning the normally nonlethal bacteria into a killer (p. 3-A).

Porter said that, although it is too early to conclude that the bacteria attacking the coral comes from humans, sewage disposal practices in the Florida Keys are being investigated as the possible cause. He said sewage dumped from cruise ships may also contribute to the problem (Nesmith 2002, 3-A; Patterson et al. 2002, 8725).

IS CORAL REEF BLEACHING A DEFENSE MECHANISM?

Andrew C. Baker wrote in *Nature* that bleaching may be a sign of a coral reef's defenses in coping with environmental change, including a warming habitat. "Coral bleaching can promote rapid response to environmental change by facilitating compensatory change in algal symbiont communities," he wrote (2001, 765). In other words, bleaching helps the corals adjust to new conditions: "Reef corals are flexible associations that can switch or shuffle symbiont communities in response to environmental change" (p. 765). There are costs, however—some symbionts that are unsuited to the new conditions do die. As described by Baker, "Bleaching is an ecological gamble in that it sacrifices short-term benefits for long-term advantage" (p. 766). Coral symbionts are, according to Baker, individually fragile, but the entire reef's flexibility gives it remarkable endurance: "[B]leaching may ultimately help reef corals to survive the recurrent and increasingly severe warming events projected by current climate models" (p. 766).

A lively debate has developed in scientific journals regarding how quickly some corals may adapt to thermal stress (Baker et al. 2004, 741; Rowan 2004, 742). Ove Hoegh-Guldberg and colleagues take issue with Andrew C. Baker's assertions that coral "bleaching favours new host-symbiont combinations that guard populations of corals against rising sea temperature" (2001, 602). Baker replied that "far fewer corals in the far-eastern Pacific Ocean died after the 1997–1998 El Niño event (20 to 26 per cent) than after the 1982–83 El Niño event (52 to 97 per cent) even though the magnitude and duration of sea-surface temperature anomaly in the region in 1997–98 exceeded those of 1982–83" (Baker, reply to Hoegh-Guldberg et al. 2001, 602).

PROTECTING CORAL REEFS

If protected from various human insults, some coral reefs may recover quickly. Coral reefs have been present for 100 million years, and they can be surprisingly resilient. "Ecologists have been surprised at how quickly reef fish populations have rebounded at successful marine parks," Hodgson said. "Two years is enough to replenish many species. Some types of corals also grow very quickly—like corn—but others take decades to reach a large size" (Hirsch 2002, 14).

Proposals have been made to establish networks of marine reserves where fishing is prohibited. By minimizing the stresses of overfishing, the corals may be able to cope with other stresses, such as global warming, according to researcher Callum M. Roberts. Roberts' study asserts that protection of ten coral reef "hot spots" around the world could save a large number of marine species. The ten reefs account for 0.017 percent of the oceans' surface area but are home to 34 percent of all species with limited ranges (Radford 2002, 12).

Roberts and his colleagues looked at eighteen areas with the greatest concentrations of species found nowhere else and selected the most vulnerable, in the Philippines, the Gulf of Guinea, the Sunda Islands in Indonesia, the southern Mascarene Islands in the Indian Ocean, eastern South Africa, the northern Indian Ocean, southern Japan, Taiwan and southern China, the Cape Verde Islands, the western Caribbean, the Red Sea, and Gulf of Aden.

Eight of the ten reefs that Roberts and colleagues would like to protect already are being dramatically altered by human activity through fishing, logging, and farming. The felling of forests, for example, means that soils are easily eroded, often depositing mud that can choke reefs. Farming releases nutrients that encourage seaweeds to grow where corals would once flourish (Radford 2002, 12).

"One of the arguments is that there is nothing we can do, it is all going to go to hell, and that coral reefs are doomed. The other argument is that we should work very hard to try and do something about protecting them," Roberts said. "The question then is how? Where are we going to focus our efforts, given that we don't have the resources to do all that we would like? We cannot save all coral reefs everywhere" (Radford 2002, 12).

WARMING THREATENS A "LIVING FOSSIL" FISH

The coelacanth, a "living fossil" fish, which has been swimming the seas for 400 million years, is being threatened by the same changes in ocean temperatures that are leading to the destruction of life-nurturing coral reef systems. "The coelacanths are vulnerable and global warming could affect them adversely," said Horst Kleinschmidt, deputy director general of South Africa's Department of Environmental Affairs and Tourism (Stoddard 2001, B-4). Previously assumed to be extinct, a colony of living coelacanth has been found in waters near South Africa, living in the reefs of Sodwana, where bleaching prompted by warming ocean temperatures is threatening the ecosystem.

WARMING SEAS AND DISEASES IN MARINE WILDLIFE

Global warming may be associated with a significant increase of unusual diseases among marine life. Many of these marine illnesses, such as *Aspergillosis*, a coral-killing fungal infection, are not yet well understood and may pose large economic problems in fisheries and even threaten human health.

Speaking at a U.S. House of Representatives Oceans Caucus Forum during October 2003, scientists associated these diseases with the discharge of sewage into waters and increased groundwater runoff. Scientists described a rise in influenza A and B in seals and other marine mammals, as well as *morbillivirus*, which is similar to feline distemper. Such diseases are common in land mammals but until now had not been noted in sea creatures. The scientists also noted an increase of the herpes virus among sea turtles and skin disorders among dolphins (Burnham 2003).

"The worst part is that we don't really have any effective tools for fighting this," said C. Drew Harvell, a professor of ecology and evolutionary biology at Cornell University (Burnham 2003). Harvell noted a growing body of evidence suggesting that global climate change is a major reason for increased marine pathogen development, disease transmission, and host susceptibility. "The level of impacts we're seeing from these diseases are new and unusual," Harvell said. "We need to figure out the human sources of these [pathogens] and stop the inputs" (2003).

As a result of warmer temperatures, harmful algal blooms have been increasing worldwide, affecting the entire food chain, said panelist Rita Colwell, president of the National Science Foundation. Harmful algal blooms, often called "red tides," result from rapid reproduction of phytoplankton in warm water. During red tides, a small number of species produce potent neurotoxins that can be transferred up the food chain and harm higher life forms, such as shellfish, fish, birds, and even humans that are exposed to the toxins (Burnham 2003).

EL NIÑO AND DEATHS OF GRAY WHALES

During 1999 and 2000, following intense El Niño conditions that sharply reduced their food supply, hundreds of gray whales floated ashore along the U.S., Canadian, and Mexican west coasts, from Alaska to Puget Sound, to San Francisco Bay and Baja California. According to a report in the *Los Angeles Times,* "The putrid carcasses became such a nuisance in 1999 and 2000 that beach communities took to towing the 35-ton cadavers out to sea or burying them with backhoes. Eskimo whalers reported harpooning 'stinky' whales that appeared to be rotting alive, too smelly even for dogs to eat" (McFarling and Weiss 2002, 1).

The whales, which weigh thirty-five to fifty tons each, spend their summers in the Bering Strait, gorging on millions of amphipods—crustaceans that live in tubes in the mud and sand on the shallow ocean floor. The whales can eat other foods, but the amphipods make up 95 percent of their diets in the Arctic. During the last fifty years, much of the ice that covers Arctic waters has melted earlier than usual. Lack of ice disrupts the food web. Even though the Arctic has been warming on average, a few locations were colder than usual. In some areas, the late spring ice has been slow to recede, keeping whales from reaching accustomed feeding grounds.

The die-off of whales largely ended after El Niño subsided, but gray whale populations plunged by more than one-third, falling from an estimated peak of 26,600 in 1998 to about 17,400 during the spring of 2002, the lowest in nearly two decades. "That's a jolting decline for a long-lived species," said Ray Highsmith, a professor of marine science at the University of Alaska at Fairbanks and an expert on the main food source for gray whales. "If the numbers are right, there's something seriously wrong" (McFarling and Weiss 2002, 1). The recovering population of whales may have been eating more than nature supplied. At the same time, nature itself, buffeted by global warming and shorter-term climate changes such as El Niño, may have been producing less of the cocktail shrimp–sized, sea-floor amphipods that are the primary food of grays (McFarling and Weiss 2002, 1).

"All of a sudden, in 1999, the bottom fell out. We went from 1,400 calves to 420. Strandings jumped from 35 to 270," said Wayne Perryman, a biologist with the Southwest Fisheries Science Center in La Jolla. "That's not a subtle signal" (McFarling and Weiss 2002, 1). The casualties didn't just include the sick, weak, young, and very old. Many of the dead animals were in the prime of their fifty-year life spans.

In the spring of 2000, veterinarian Frances Gulland of Sausalito's Marine Mammal Center conducted full necropsies on three animals and found as many distinct causes of death: viral encephalitis, the biotoxin domoic acid, and parasitic abscesses. "All of those could have initially started as malnutrition," Gulland said. "The real question is, why were they so malnourished? Why did they get whatever caused them to die?" (McFarling and Weiss 2002, 1). Both the living and dead animals were so skinny that their ribs stuck out. Their scrawniness was visible even from aerial photographs.

WARMING WATERS CHOKE LIFE OUT OF LAKE TANGANYIKA

Two independent teams of scientists studying central Africa's Lake Tanganyika, Africa's second-largest body of fresh water, have found a microcosm of crisis vis-à-vis global warming. The scientists have found that warming at the lake's surface has impaired mixing of nutrients, reducing its population of fish. These reductions have affected the local economy as fishing yields fell by a third or more during thirty years, with more declines anticipated. Heretofore, Lake Tanganyika's fish had supplied 25–40 percent of the protein consumed by neighboring peoples in parts of Burundi, Tanzania, Zambia, and the Democratic Republic of Congo.

Lake Tanganyika is a tropical body of water that experiences relatively high temperatures year-round, so the scientists were rather surprised to discover that further warming affected its nutrient balance so significantly. Like other deep-water lakes, however, Tanganyika relies on temperature differences at various depths to mix water and nutrients. Such mixing is very critical in tropical lakes with sharp temperature gradients that stratify layers of water, with warm, less-dense layers on top of nutrient-rich waters below.

"Climate warming is diminishing productivity in Lake Tanganyika," Catherine M. O'Reilly from the University of Arizona, who led one of the study teams, and colleagues wrote. "In parallel with regional warming patterns since the beginning of the twentieth century," they continued, "a rise in surface-water temperature has increased the stability of the water column" (O'Reilly et al. 2003, 766). A regional decrease in average wind speed over the lake also has contributed to reduced mixing of the 1,470-meter deep lake, "decreasing deep-water nutrient upwelling and entrainment into surface waters" (p. 766). Fish yields have declined roughly 30 percent, the scientists wrote, an example "that the impact of regional effects of global climate change on aquatic ecosystem functions can be larger than that of local anthropogenic activity or over-fishing" (p. 766). Lake Tanganyika is especially vulnerable because year-round tropical temperatures accelerate biological processes, "and new nutrient inputs from the atmosphere or rock weathering cannot keep up with the high rates of algal photosynthesis and decomposition" (Verschuren 2003, 731).

Lake Tanganyika is the second-deepest lake in the world and the second-richest in terms of biological diversity; it has at least 350 species of fish, with new ones being discovered regularly. Nutrient mixing has been vital for its biodiversity (Connor 2003). Piet Verburg of the University of Waterloo in Canada, who led the other study on Lake Tanganyika, found that warmer temperatures and less windy weather in the region have been starving the lake of essential salts that contain nitrogen and sulfur (Verburg, Hecky, and Kling 2003, 505–507). Verburg and colleagues utilized profiles of temperature changes in the lake between 110 and 800 meters and found that degree of temperature stratification has tripled since 1913.

O'Reilly and colleagues, writing in *Nature*, suggested that the lake's productivity, measured by the amount of photosynthesis, has diminished by 20 percent, which could easily account for the 30 percent decline in fish yields. The scientists said that climate change, rather than over-fishing, was mainly responsible for the collapse in Tanganyika's fish stocks. With additional warming, fish populations in the lake are expected by the scientists to decline even further (Connor 2003). "The human implications of such subtle, but progressive, environmental changes are potentially dire in this densely populated region of the world, where large lakes are essential natural resources for regional economies," the scientists said.

Dirk Verschuren, a freshwater biologist at Ghent University in Belgium, said that both studies could explain why sardine catches in Lake Tanganyika have declined between 30 and 50 percent since the late 1970s (Verschuren 2003, 731). "Since overexploitation is at most a local problem on some fishing grounds, the principal cause of this decline has remained unknown," Verschuren said. "Taken together ... the data in the two papers provide strong evidence that the effect of global climate change on regional temperature has had a greater impact on Lake Tanganyika than have local human activities. Their combined evidence covers all the important links in the chain of cause and effect between climate warming and the declining fishery (Connor 2003).

INCREASING POPULATIONS AND POTENCY OF JELLYFISH

Some sea species thrive on conditions that kill others. For example, jellyfish seem to have a biological affinity for some forms of human pollution as well as warmer habitats. Jellyfish populations have been increasing rapidly in many parts of the world. In some areas the increase appears to be part of a natural cycle (jellyfish populations also are declining in a few other areas) (Pohl 2002, F-3). By the summer of 2004, reports indicated that jellyfish populations were on the rise in Puget Sound, the Bering Strait, and the harbors of Tokyo and Boston. "Smacks" or swarms of jellyfish shut down fisheries in Narragansett Bay, parts of the Gulf of Alaska, and sections of the Black Sea. In the Philippines, fifty tons of jellyfish shut down a power plant, provoking blackouts, when they were sucked into its cooling system (Carpenter

2004, 68). In late July 2003, thousands of barrel jellyfish and moon jellyfish washed up on the coast of southern Wales.

"Jellies are a pretty good group of animals to track coastal ecosystems," said Monty Graham, a scientist at the University of South Alabama. "When you start to see jellyfish numbers grow and grow, that usually indicates a stressed system" (Pohl 2002, F-3). Those stresses include increased water temperature, a rise in nutrients (from fertilizers and sewage), and depleted stocks of other fish, often caused by overfishing, which removes the jellyfish's competitors. All of these changes are usually human-caused, according to Graham.

Otto Pohl of the *New York Times* described the increasing numbers and toxic potency of jellyfish in some areas of Australia that are popular with swimmers:

When Robert King climbed back on the boat after snorkeling off the Great Barrier Reef…on March 31 [2002], he knew something was wrong. "I don't feel so good," he said, rubbing his chest. He had been stung by a jellyfish, and his condition deteriorated rapidly. By the time [an] emergency helicopter arrived, he was screaming in agony; a few hours later he was in a coma, eyes frozen wide, bleeding into his brain. He never regained consciousness. King, 44, from Columbus, Ohio, was the second person in Australia to die this year from the sting of a species of jellyfish, *Carukia barnesi*, found only in Australia and never before known to be fatal. More than 200 other victims went to hospitals, several times the number in a normal summer season here….Jellyfish release millions of microscopic harpoons when touched, shooting tiny hypodermic needles into a victim's skin. They are lined with barbs and filled with venom, and they often linger painfully in the skin for months after the toxin has worn off. (2002, F-3)

Speaking of jellyfish stings in Australia, Jamie Seymour, a jellyfish expert at James Cook University, commented, "This year [2002] [was] incredibly abnormal" (Pohl 2002, F-3). Seymour believes that strong, unusual wind patterns helped blow the jellyfish toward the shore, where they flourished in unseasonably warm waters. Seymour, who has analyzed the venom from each sting that receives hospital treatment in the Great Barrier Reef region for years, had never seen the type of venom that killed the two tourists in 2002. In the Gulf of Mexico, according to the report in the *New York Times*, shrimp fishermen were struggling with rising numbers of jellyfish that filled their nets with slimy gelatin, ruining their catch (2002, F-3).

On Waikiki Beach in Hawaii, a lifeguard, Landy Blair, counted jellyfish stings on more than 900 people during a single day in 2002, about 1 percent of which sent victims to hospitals. Blair has been keeping track of jellyfish populations near the beach since 1991. The problem has grown steadily worse, he said, adding, "We have seen the highest numbers ever over the past year" (Pohl 2002, F-3). On the beaches near Auckland, New Zealand, half a dozen sting victims have required hospitalization, said Robert Ferguson of Surf Life Saving New Zealand, a lifeguards' organization. "It's the first time I've ever heard of victims needing hospital care," Ferguson said. "This is a new type of jellyfish with stings that are much more severe, much harsher" (2002, F-3).

At about the same time that increasingly potent jellyfish were being observed in the South Pacific, a report appeared in the *Boston Globe* describing a massive infestation of jellyfish in Narragansett Bay and Long Island Sound. A group of fishermen who expected "an array of marine life in their nets . . . got jellyfish, nothing but jellyfish; jellyfish so plentiful that the gelatinous organisms came up dangling through the net like slimy icicles. And with each haul came more" (Arnold 2002, C-1).

"Eventually it seemed that our deck was coated with vaseline," said Captain Eric Pfirrmann, who works for Save the Bay, a group whose members engage in environmental issues related to Rhode Island's Narragansett Bay. He piloted a research vessel that had taken several high-school teachers on a marine field trip. "I've seen blooms like this before," Pfirrmann said, "but never so early in the summer." The culprit was a non-stinging invertebrate about the size and shape of a tulip blossom and commonly known as the combjelly. These jellyfish, along with sea squirts (an entirely different organism) were taking over Long Island Sound, thriving in large part because water temperatures have risen about 3 degrees F over the past two decades, according to scientists (Arnold 2002, C-1).

The rapid increase in jellyfish has nearly wiped out the winter flounder population in Narragansett Bay at the same time that the non-native sea squirt has overrun local oysters and blue mussels. Both are favored by warmer water temperatures and human pollution. "There is evidence of jellyfish explosions around the world that appear related to the adverse impact of human activities, and those include global warming," said Sarah Chasis, the senior attorney for the New York City–based Natural Resources Defense Council (Arnold 2002, C-1).

The *Boston Globe* report said that "historically, Narragansett Bay was the northern limit of the combjelly, whose domain extends as far south as Argentina. During November of 2000 combjellies were documented for the first time in Boston Harbor, although in numbers too sparse at that time to affect the harbor's ecology. In Narragansett Bay, according to Barbara Sullivan, an oceanographer from the University of Rhode Island who has been studying the jellyfish infestation, warmer water is changing the rules of who eats whom" (Arnold 2002, C-1). The combjelly's reproductive cycle adjusts to the warmth of the water in which it lives. Populations usually explode during the warmth of late summer and early autumn. Narragansett Bay has warmed an average of 3.4 degrees F during the past twenty years, whereas between the late 1970s to 2001 the average temperature of Long Island Sound during the first three months of each year has increased about 8 percent, from 37.4–40.2 degrees F (p. C-1).

As waters have warmed, the combjelly has reproduced and "bloomed" four months earlier, which enables it to gobble up the eggs and larvae deposited in the spring by spawning fish to the point of nearly replacing them. "We have seen areas of the bay where these things have cleaned out everything edible floating in the water column," Chasis said. The winter flounder population has nearly vanished. In 1982, approximately 4,200 metric tons of flounder were landed in Rhode Island. In 2000, the most recent year for which statistics were available, approximately 600 tons of flounder were caught. "Comb-jellies do indeed eat many winter flounder larvae," said Tim Lynch, a marine biologist with the state Division of Fish and Wildlife, "but I do not believe they are the sole cause of the population decline in Narragansett Bay." Factors such as overfishing and water quality probably also play a role, Lynch said (Arnold 2002, C-1).

GIANT SQUID BENEFIT FROM WARMING

Another species that seems to benefit from a warming environment is the giant squid. Squids have reproduced with much success, increasing their physical size and numbers to a point where, by the end of the twentieth century, they had overtaken humans in terms of total planetwide biomass, according to some estimates. Human overfishing of other species has favored the squid (reducing competition for food), along with a warmer habitat. Australian scientist George Jackson said he believed that global warming is causing squid to grow larger ("Giant Squid" 2002).

A report in the Australian science journal *Australasian Science* said that most marine researchers are now in agreement that warming habitats have given cephalopods an advantage not available to any other large sea creature. As a result, they have flourished. Their numbers have been increasing, along with their body size (Benson 2002, 4). In July 2002, according to a report by Agence France Presse, the body of a giant squid weighing about 200 kilograms (440 pounds) washed up on a beach in the Australian state of Tasmania. Within days of that event, hundreds of dead squid washed ashore on the coast of California.

George Jackson, with the Institute of Antarctic and Southern Ocean Studies in Tasmania, said squid thrive during environmental disasters such as global warming. The animal eats anything that comes its way, breeds whenever possible, and keeps growing.

> This trend has been suggested to be due both to the removal of cephalopod predators such as toothed whales and tuna and an increase of cephalopods due to removal of finfish competitors. . . . The fascinating thing about squid is that they're short-lived. I haven't found any tropical squid in Australia older than 200 days. . . . Many of the species have exponential growth, particularly during the juvenile stage so if you increase the water temperature by even a degree it has a tremendous snowballing effect of rapidly increasing their growth rate and their ultimate body size. They get much bigger and they can mature earlier and it just accelerates everything. (Benson 2002, 4)

REFERENCES:
PART IV.
WARMING SEAS

Aitken, Mike. "St. Andrews Stymied by Natural Hazard." *The Scotsman*, April 18, 2001, 20.

Alley, Richard B., Peter U. Clark, Philippe Huybrechts, and Ian Joughin. "Ice-Sheet and Sea-Level Changes." *Science* 310 (October 21, 2005): 456–460.

Amor, Adlai. "Report Warns of Growing Destruction of World's Coastal Areas." World Resources Institute, April 17, 2001. www.dooleyonline.net/media_preview/index.cfm.

Arnold, David. "Global Warming Lends Power to a Jellyfish in Narragansett Bay and Long Island Sound: Non-Native Species Are Taking Over." *Boston Globe*, July 2, 2002, C-1.

Arthur, Charles. "Super El Niño Could Turn Amazon into Dustbowl: British Association for the Advancement of Science." *London Independent*, September 9, 2003, 6.

———. "Temperature Rise Kills 90 Per Cent of Ocean's Surface Coral." *London Independent*, September 18, 2003. (Lexis).

Baker, Andrew C. "Reef Corals Bleach to Survive Change." *Nature* 411 (June 14, 2001): 765–766.

———, Craig J. Starger, Tim R. McClanahan, and Peter W. Glynn. "Corals' Adaptive Response to Climate Change." *Nature* 430 (August 12, 2004): 741.

Barbeliuk, Anne. "Warmer Globe Choking Ocean." *Hobart (Australia) Mercury*, March 16, 2002. (Lexis).

Barkham, Patrick. "Going Down: Tuvalu, a Nation of Nine Islands—Specks in the South Pacific—Is in Danger of Vanishing, a Victim of Global Warming. As Their Homeland Is Battered by Ferocious Cyclones and Slowly Submerges under the Encroaching Sea, What Will Become of the Islanders?" *London Guardian*, February 16, 2002, 24.

Barnett, Tim P., David W. Pierce, Krishna M. AchutaRao, Peter J. Gleckler, Benjamin D. Santer, Jonathan M. Gregory, et al. "Penetration of Human-Induced Warming into the World's Oceans." *Science* 309 (July 8, 2005): 284–287.

———, David W. Pierce, and Reiner Schnur. "Detection of Anthropogenic Climate Change in the World's Oceans." *Science* 292 (April 13, 2001): 270–274.

Beaugrand, Gregory, Keith M. Brander, J. Alistair Lindley, Sami Souissi, and Philip C. Reid. "Plankton Effect on Cod Recruitment in the North Sea." *Nature* 426 (December 11, 2003): 661–664.

———, Philip C. Reid, Frédéric Ibañez, J. Alistair Lindley, and Martin Edwards. "Reorganization of North Atlantic Marine Copepod Biodiversity and Climate." *Science* 296 (May 31, 2002): 1692–1694.

Bellwood, D. R., T. P. Hughes, C. Folke, and M. Nystrom. "Confronting the Coral Reef Crisis." *Nature* 429 (June 24, 2004): 827–833.

Benson, Simon. "Giant Squid 'Taking Over the World.' " *Sydney Daily Telegraph*, July 31, 2002, 4.

Bond, G., W. Showers, M. Cheseby, R. Lotti, P. Almasi, P. deMenocal, et al. "A Pervasive, Millenial-Scale in North Atlantic Holocene and Glacial Climates." *Science* 278 (1997): 1257–1266.

Borenstein, Seth. "Scientists Worry about Evidence of Melting Arctic Ice." Knight-Ridder News Service, *Seattle Times*, February 18, 2005, A-6.

Bourne, Joel K., Jr. "The Big Uneasy." *National Geographic*, October 2004, 88–105.

Boyd, Robert S. "Rising Tides Raises Questions: Satellites Will Provide Exact Measurements." Knight-Ridder Newspapers, *Pittsburgh Post-Gazette*, December 9, 2001, A-3.

Breed, Allen G. "New Orleans Evacuation Picking Up Steam, but Help Comes Too Late for Untold Number." Associated Press, September 3, 2005. (Lexis).

Broecker, Wallace S. "Are We Headed for a Thermohaline Catastrophe?" In *Geological Perspectives of Global Climate Change*, Studies in Geology #17, ed. Lee C. Gerhard, William E. Harrison, and Bernold M. Hanson, 83–95. Tulsa, OK: American Association of Petroleum Geologists, 2001.

———. "Thermohaline Circulation: The Achilles Heel of Our Climate System: Will Man-Made CO_2 Upset the Current Balance?" *Science* 278 (1997): 1582–1588.

———. "Unpleasant Surprises in the Greenhouse?" *Nature* 328 (1987): 123–126.

Brown, Paul, and Tony Sutton. "Global Warming Brings New Cash Crop to West Country as Rising Water Temperatures Allow Valuable Shellfish to Thrive." *London Guardian*, December 10, 2002, 8.

Browne, Anthony. "Canute Was Right! Time to Give Up the Coast." *London Times,* October 11, 2002, 8.

Bryden, Harry L., Hannah R. Longworth, and Stuart A. Cunningham. "Slowing of the Atlantic Meridional Overturning Circulation at 25° North." *Nature* 438 (December 1, 2005): 655–657.

Bunting, Madeleine. "Confronting the Perils of Global Warming in a Vanishing Landscape: As Vital Talks Begin at The Hague, Millions Are Already Suffering the Consequences of Climate Change." *London Guardian*, November 14, 2000, 1.

Burnham, Michael. "Scientists Link Global Warming with Increasing Marine Diseases." Greenwire, October 7, 2003. (Lexis).

Calamai, Peter. "Atlantic Water Changing: Scientists." *Toronto Star*, June 21, 2001, A-18.

Caldeira, Ken, and Michael E. Wickett. "Oceanography: Anthropogenic Carbon and Ocean pH." *Nature* 425 (September 25, 2003): 365.

Carpenter, Betsy. "Feeling the Sting: Warming Oceans, Depleted Fish Stocks, Dirty Water—They Set the Stage for a Jellyfish Invasion." *U.S. News and World Report*, August 16, 2004, 68–69.

Chen, Junye, Barbara E. Carlson, and Anthony D. Del Genio. "Evidence for Strengthening of the Tropical General Circulation in the 1990s." *Science* 295 (February 1, 2002): 838–841.

Clark, D. A., S. C. Piper, C. D. Keeling, and D. B. Clark. "Tropical Rain Forest Tree Growth and Atmospheric Carbon Dynamics Linked to Interannual Temperature Variation during 1984–2000." *Proceedings of the National Academy of Sciences* 100 (10) (May 13, 2003): 5852–5857.

Clark, Peter U., A. Marshall McCabe, Alan C. Mix, and Andrew J. Weaver. "Rapid Rise of Sea Level 19,000 Years Ago and Its Global Implications." *Science* 304 (May 21, 2004): 1141–1144.

———, N. G. Pisias, T. F. Stocker, and A. J. Weaver. "The Role of the Thermohaline Circulation in Abrupt Climate Change." *Nature* 415 (February 21, 2002): 863–868.

Clover, Charles. "Global Warming 'Is Driving Fish North.'" *London Daily Telegraph*, May 31, 2002, 14.

"Coastal Gulf States Are Sinking." Environment News Service, April 21, 2003. http://ens-news.com/ens/apr2003/2003-04-21-09.asp#anchor4.

Cole, Julia. "A Slow Dance for El Niño." *Science* 291 (February 23, 2001): 1496–1497.

Connor, Steve. "Britain Could Become as Cold as Moscow." *London Independent*, June 21, 2001, 14.

————. "El Niños' Rise May Be Linked to Pacific Current Slowdown." *London Independent*, February 7, 2002.

————. "Global Warming Is Choking the Life Out of Lake Tanganyika." *London Independent*, August 14, 2003. (Lexis).

————. "Strangers in the Seas: Exotic Marine Species Are Turning Up Unexpectedly in the Cold Waters of the North Atlantic." *London Independent*, August 5, 2002, 12–13.

Cooke, Robert. "Scientists: Pacific Slower at Surface: Data May Help Explain Trend in El Niño Events." *Newsday*, February 7, 2002, A-46.

————. "Waters Reflect Weather Trend: Study Finds Warming Effects." *Newsday*, December 18, 2003, A-2.

Cowen, Robert C. "Into the Cold? Slowing Ocean Circulation Could Presage Dramatic—and Chilly—Climate Change." *Christian Science Monitor*, September 26, 2002, 14.

Cramb, Auslan. "Highland River Salmon 'On Verge of Extinction.' " *London Daily Telegraph*, July 15, 2002, 7.

Crenson, Matt. "Louisiana Sinking: One State's Environmental Nightmare Could Become Common Problem." Associated Press, August 10, 2002. (Lexis).

Curry, Ruth, Bob Dickson, and Igor Yashayaev. "A Change in the Freshwater Balance of the Atlantic Ocean over the Past Four Decades." *Nature* 426 (December 18, 2003): 826–829.

————, and Cecilie Mauritzen. "Dilution of the Northern North Atlantic Ocean in Recent Decades." *Science* 308 (June 17, 2005): 1772–1774.

Davidson, Keay. "Going to Depths for Evidence of Global Warming: Heating Trend in North Pacific Baffles Researchers." *San Francisco Chronicle*, March 1, 2004, A-4.

Dean, Cornelia. "Louisiana's Marshes Fight for Their Lives." *New York Times*, November 15, 2005. www.nytimes.com/2005/11/15/science/earth/15marsh.html.

Dickson, B., I. Yashayaev, J. Meincke, B. Turrell, S. Dye, and J. Holfort. "Rapid Freshening of the Deep North Atlantic Ocean over the Past Four Decades." *Nature* 416 (April 25, 2002): 832–836.

Erbacher, Jochen, Brian T. Huber, Richard D. Norris, and Molly Markey. "Increased Thermohaline Stratification as a Possible Cause for an Ocean Anoxic Event in the Cretaceous Period." *Nature* 409 (January 18, 2001): 325–327.

Evans-Pritchard, Ambrose. "Dutch Have Only Years before Rising Seas Reclaim Land: Dikes No Match against Global Warming Effects." *London Daily Telegraph* in *Ottawa Citizen*, September 8, 2004, A-6.

"Fear as Water Bleaches Reef." *Sydney (Australia) Daily Telegraph*, January 9, 2002, 12.

Feely, Richard A., Christopher L. Sabine, Kitack Lee, Will Berelson, Joanie Kleypas, Victoria J. Fabry, et al. "Impact of Anthropogenic CO_2 on the $CaCO_3$ System in the Oceans." *Science* 305 (July 16, 2004): 362–367.

Field, Michael. "Dying Pacific Breadfruit New Sign of Looming Disaster." Agence France Presse, December 1, 2002. (Lexis).

"Fished to the Point of Ruin, North Sea Cod Stocks So Low as to Spell Disaster." *Glasgow Herald*, November 7, 2000, 18.

Freeman, James, and Eleanor Cowie. "Pollutants Threaten the Great Barrier Reef." *Glasgow Herald*, January 25, 2002, 7.

Freemantle, Tony. "Global Warming Likely to Hit Texas: Scientists Say Temperature Rise Will Change Rainfall, Gulf Coast Region." *Houston Chronicle*, October 24, 2001, 32.

Fukasawa, Masao, Howard Freeland, Ron Perkin, Tomowo Watanabe, Hiroshi Uchida, and Ayako Nishina. "Bottom Water Warming in the North Pacific Ocean." *Nature* 427 (February 26, 2004): 825–827.

Gagosian, Robert B. "Abrupt Climate Change: Should We Be Worried?" Woods Hole Oceanographic Institution, January 27, 2003.

Gardner, Toby A., Isabelle M. Côté, Jennifer A. Gill, Alastair Grant, and Andrew R. Watkinson. "Long-Term Region-Wide Declines in Caribbean Corals." *Science* 301 (August 15, 2003): 958–960. Posted online July 18, 2003, www.scienceexpress.org.

"Giant Squid Film Team Makes Spectacular Catch." Agence France Presse, September 14, 2002.

Gille, Sarah T. "Warming of the Southern Ocean since the 1950s." *Science* 295 (February 15, 2002): 1275–1277.

"Global Warming Blamed for Rising Sea Levels." Associated Press in *Omaha World-Herald*, November 25, 2001, 20-A.

"Global-Warming Signal from the Ocean." *Dallas Morning News* in *New Orleans Times-Picayune*, December 31, 2000, 4.

Goes, Joaquim I., Prasad G. Thoppil, Helga do R. Gomes, and John T. Fasullo. "Warming of the Eurasian Landmass Is Making the Arabian Sea More Productive." *Science* 308 (April 22, 2005): 545–547.

Great Barrier, Simon. "Giant Squid 'Taking over the World.'" *Sydney Daily Telegraph*, July 31, 2002, 4.

"Great Barrier Reef Is Springing Back to Life." *Western Daily Press (Australia)*, December 18, 2002, 11.

Gregg, Watson W., and Margarita E. Conkright. "Decadal Changes in Global Ocean Chlorophyll. *Geophysical Research Letters* 29 (15) (2002): 1–4. doi: 10.1029/2002GL014689.

Gregory, Angela. "Fear of Rising Seas Drives More Tuvaluans to New Zealand." *New Zealand Herald*, February 19, 2003. (Lexis).

Häkkinen, Sirpa, and Peter B. Rhines. "Decline of Subpolar North Atlantic Circulation during the 1990s." *Science* 304 (April 23, 2004): 555–559.

Hale, Ellen. "Seas Create Real Water Hazard: Changing Climate at Root of Erosion That's Putting Links Courses in Jeopardy." *USA Today*, July 18, 2001, 3-C.

Hall, Carl T. "Ocean Tells the Story: Earth Is Heating Up; Human Activity, Not Variables in Nature, Cited as Culprit." *San Francisco Chronicle*, April 29, 2005, A-1.

Hansen, Bogi, William R. Turrell, and Svein Sterhus. "Decreasing Overflow from the Nordic Seas into the Atlantic Ocean through the Faroe Bank Channel since 1950." *Nature* 411 (June 21, 2001): 927–930.

Hansen, James E. "Defusing the Global Warming Time Bomb." *Scientific American* 290 (3) (March 2004): 68–77.

———, Larissa Nazarenko, Reto Ruedy, Makiko Sato, Josh Willis, Anthony Del Genio, et al. "Earth's Energy Imbalance: Confirmation and Implications." *Science* 308 (June 3, 2005): 1431–1435.

Hartmann, Dennis L. "Climate Change: Tropical Surprises." *Science* 295 (February 1, 2002): 811–812.

Healy, Patrick. "Warming Waters: Lobstermen on Cape Cod Blame Light Hauls on Higher Ocean Temperatures." *Boston Globe*, August 30, 2002, B-1.

"Heavy Rains Threaten Flood-Prone Venus." *Singapore Straits Times*, June 8, 2002. (Lexis).

Hirsch, Jerry. "Damage to Coral Reefs Mounts, Study Says: Broad Survey Cites Human Causes such as Over-Fishing and Pollution; Reefs Are a Key Indicator of the Health of Oceans, Scientists Say." *Los Angeles Times*, August 26, 2002, 14.

Hoegh-Guldberg, Ove, Ross J. Jones, Selina Ward, and William K. Loh. "Is Coral Bleaching Really Adaptive?" *Nature* 415 (February 7, 2001): 601–602.

Hollingsworth, Jan. "Global Warming Studies Put Heat on State: Tampa Bay Area Labeled Extremely Vulnerable." *Tampa Tribune*, October 24, 2001, 1.

Holly, Chris. "Sea-Level Rise Seen as Key Global Warming Threat." *The Energy Daily* 32 (36) (February 25, 2004). (Lexis).

Houghton, J. T., Y. Ding, D. J. Griggs, M. Noguer, P. J. van der Linden, X. Dai, et al. *Climate Change 2001: The Scientific Basis. Contribution of Working Group I to the Third Assessment Report of the Intergovernmental Panel on Climate Change.* Cambridge, U.K.: Cambridge University Press, 2001.

Huber, Matthew, and Rodrigo Caballero. "Eocene El Niño: Evidence for Robust Tropical Dynamics in the 'Hothouse.'" *Science* 299 (February 7, 2003): 877–881.

Hughes, T. P., A. H. Baird, D. R. Bellwood, M. Card, S. R. Connolly, C. Folke, et al. "Climate Change, Human Impacts, and the Resilience of Coral Reefs." *Science* 31 (August 15, 2003): 929–933.

Huq, Saleemul. "Climate Change and Bangladesh." *Science* 294 (November 23, 2001): 1617.

Inkley, D. B., M. G. Anderson, A. R. Blaustein, V. R. Burkett, B. Felzer, B. Griffith, et al. *Global Climate Change and Wildlife in North America.* Washington, D.C.: The Wildlife Society, 2004. www.nwf.org/news.

Jenkyns, Hugh C., Astrid Forster, Stefan Schouten, and Jaap S. Sinninghe Damste. "High Temperatures in the Late Cretaceous Artic Ocean." *Nature* 432 (December 16, 2004): 888–892.

Jordan, Steve. "Wary Insurers Suspect Climate Change." *Omaha World-Herald*, September 1, 2005, 7-A.

Joughin, I., W. Abdalati, and M. Fahnestock. "Large Fluctuations in Speed on Greenland's Jakobshavn Isbrae Glacier." *Nature* 432 (December 2, 2004): 608–610.

"Jumbo Squid Has a Message for Us: Changing Global Patterns Are Going to Bring Different Species into Our Waters." *Victoria (British Columbia) Times-Colonist*, October 8, 2004, A-10.

Kaiser, Jocelyn. "Reproductive Failure Threatens Bird Colonies on North Sea Coast." *Science* 305 (August 20, 2004): 1090.

Kelleher, Lynne. "Look Who's Here: Tropical Fish Warming to Waters around Ireland." *London Sunday Mirror*, October 20, 2002, 15.

"Kelp Points to Worrying Sea Change." *Canberra Times*, August 30, 2004, A-8.

Kerr, Richard A. "Climate Change: Sea Change in the Atlantic." *Science* 303 (January 2, 2004): 35.

———. "El Niño or La Nina? The Past Hints at the Future." *Science* 309 (July 29, 2005): 687.

———. "European Climate: Mild Winters Mostly Hot Air, Not Gulf Stream." *Science* 297 (September 27, 2002): 2202.

———. "Greenhouse Warming Passes One More Test." *Science* 292 (April 13, 2001): 193.

———. "Is Katrina a Harbinger of Still More Powerful Hurricanes?" *Science* 309 (September 16, 2005): 1807.

————. "Rising Global Temperature, Rising Uncertainty." *Science* 292 (April 13, 2001): 192–194.

Knorr, Gregory, and Gerrit Lohmann. "Southern Ocean Origin for the Resumption of Atlantic Thermohaline Circulation during Deglaciation." *Nature* 424 (July 31, 2003): 532–536.

Knutti, R., J. Fluckiger, T. F. Stocker, and A. Timmermann. "Strong Hemispheric Coupling of Glacial Climate through Freshwater Discharge and Ocean Circulation." *Nature* 430 (August 19, 2004): 851–856.

Lean, Geoffrey. "Quarter of World's Corals Destroyed." *London Independent*, January 7, 2001, 7.

Leggett, Jeremy. *The Carbon War: Global Warming and the End of the Oil Era*. New York: Routledge, 2001.

Levitus, Sydney, John I. Antonov, Julian Wang, Thomas L. Delworth, Keith W. Dixon, and Anthony J. Broccoli. "Anthropogenic Warming of Earth's Climate System." *Science* 292 (April 13, 2001): 267–270.

Lynas, Mark. *High Tide: The Truth about Our Climate Crisis*. New York: Picador/St. Martins, 2004.

Macdougall, Doug. *Frozen Earth: The Once and Future Story of Ice Ages*. Berkeley: University of California Press, 2004.

Maslanik, J. A., M. C. Serreze, and T. Agnew. "On the Record Reduction in 1998 Western Arctic Sea Ice Cover." *Geophysical Research Letters* 26 (13) (1999): 1905–1912.

McCarthy, Michael. "Climate Change Provides Exotic Sea Life with a Warm Welcome to Britain." *London Independent*, January 24, 2002, 13.

————. " 'Rainforests of the Sea' Ravaged: Over-Fishing and Pollution Kill 80 Per Cent of Coral on Caribbean Reefs." *London Independent*, July 18, 2003, 3.

McCord, Joel. "Marshes in Decay Haunt the Bay." *Baltimore Sun*, December 6, 2000, 1-B.

McFadden, Robert D. "New Orleans Begins a Search for Its Dead; Violence Persists." *New York Times*, September 5, 2005.

McFarling, Usha Lee. "Studies Point to Human Role in Global Warming." *Los Angeles Times*, April 13, 2001, A-1.

————, and Kenneth R. Weiss. "A Whale of a Food Shortage." *Los Angeles Times*, June 24, 2002, 1.

McKie, Robin. "Dying Seas Threaten Several Species: Global Warming Could Be Tearing Apart the Delicate Marine Food Chain, Spelling Doom for Everything from Zooplankton to Dolphins." *London Observer*, December 2, 2001, 14.

McPhaden, Michael J., and Dongxiao Zhang. "Slowdown of the Meridional Overturning Circulation in the Upper Pacific Ocean." *Nature* 415 (February 7, 2002): 603–608.

Meier, Mark F., and Mark B. Dyurgerov. "Sea-Level Changes: How Alaska Affects the World." *Science* 297 (July 19, 2002): 350–351.

Miles, Paul. "Fiji's Coral Reefs Are Being Ruined by Bleaching." *London Daily Telegraph*, June 2, 2001, 4.

Miller, Laury, and Bruce C. Douglas. "Mass and Volume Contributions to Twentieth-Century Global Sea Level Rise." *Nature* 428 (March 25, 2004): 406–408.

Milmo, Cahal, and Elizabeth Nash. "Fish Farms Push Atlantic Salmon towards Extinction." *London Independent*, June 1, 2001, 11.

Nesmith, Jeff. "Sewage Off Keys Cripples Coral: Bacteria Causes Deadly Disease." *Atlanta Journal and Constitution*, June 18, 2002, 3-A.

"New Wave of Bleaching Hits Coral Reefs Worldwide." Environment News Service, October 29, 2002. http://ens-news.com/ens/oct2002/2002-10-29-19.asp#anchor1.

Northrop, Michael. "Adapting to Warming: The United States Can Do It, but Europe Can't." *Washington Post*, December 16, 2002, A-25.

Nosengo, Niccola. "Venice Floods: Save Our City!" *Nature* 424 (August 7, 2003): 608–609.

Nussbaum, Alex. "The Coming Tide: Rise in Sea Level Likely to Increase N.J. Floods." *Bergen County (New Jersey) Record*, September 4, 2002, A-1.

Nuttall, Nick. "Coral Reefs 'On the Edge of Disaster.'" *London Times*, October 25, 2000. (Lexis).

———. "Global Warming Boosts El Niño." *London Times*, October 26, 2000. (Lexis).

"Ocean Temperatures Reach Record Highs." Associated Press, September 9, 2002. (Lexis).

O'Harra, Doug. "Marine Parasite Infects Yukon River King Salmon: Fish Are Left Inedible; Scientists Study Overall Impacts." *Anchorage Daily News*, January 28, 2004, A-1.

O'Reilly, Catherine M., Simone R. Alin, Pierre-Denis Plisnier, Andrew S. Cohen, and Brent A. McKee. "Climate Change Decreases Aquatic Ecosystem Productivity of Lake Tanganyika, Africa." *Nature* 424 (August 14, 2003): 766–768.

Orr, James C., Victoria J. Fabry, Olivier Aumont, Laurent Bopp, Scott C. Doney, Richard A. Feely, Anand Gnanadesikan, Nicolas Gruber, Akio Ishida, Fortunat Joos, Robert M. Key, Keith Lindsay, Ernst Maier-Reimer, Richard Matear, Patrick Monfray, Anne Mouchet, Raymond G. Najjar, Gian-Kasper

Plattner, Keith B. Rodgers, Christopher L. Sabine, Jorge L. Sarmiento, Reiner Schlitzer, Richard D. Slater, Ian J. Totterdell, Marie-France Weirig, Yasuhiro Yamanaka, and Andrew Yooi. "Anthropogeenic Ocean Acidification over the Twenty-first Century and Its Impact on Calcifying Organisms." *Nature* 437 (September 29, 2005): 681–686.

"Over 80 Per Cent of Indonesia's Coral Reefs under Threat." *Jakarta Post*, September 13, 2001. (Lexis).

"Pacific Too Hot for Corals of World's Largest Reef." Environment News Service, May 23, 2002. http://ens-news.com/ens/may2002/2002-05-23-01.asp.

Patterson, Kathryn L., James W. Porter, Kim B. Ritchie, Shawn W. Polson, Erich Mueller, Esther C. Peters, et al. "The Etiology of White Pox, a Lethal Disease of the Caribbean Elkhorn Coral, *Acropora palmata*." *Proceedings of the National Academy of Sciences* 99 (13) (June 25, 2002): 8725–8730.

Pearce, Fred. "Failing Ocean Current Raises Fears of Mini Ice Age." New Scientist.com News Service, November 30, 2005. www.newscientist.com/article.ns?id=dn8398.

Pelton, Tom. "New Maps Highlight Vanishing E. Shore: Technology Provides a Stark Forecast of the Combined Effect of Rising Sea Level and Sinking Land along the Bay." *Baltimore Sun*, July 30, 2004, 1-A.

Perlman, David. "Decline in Oceans' Phytoplankton Alarms Scientists: Experts Pondering Whether Reduction of Marine Plant Life Is Linked to Warming of the Seas." *San Francisco Chronicle*, October 6, 2003, A-6.

Peterson, Bruce J., Robert M. Holmes, James W. McClelland, Charles J. Vorosmarty, Richard B. Lammers, Alexander I. Shiklomanov, et al. "Increasing River Discharge to the Arctic Ocean." *Science* 298 (December 13, 2002): 2171–2173.

Petrillo, Lisa. "Turning the Tide in Venice." Copley News Service, April 28, 2003. (Lexis).

Pilkey, Orrin H., and Andrew G. Cooper. "Society and Sea Level Rise." *Science* 303 (March 19, 2004): 1781–1782.

Pittman, Craig. "Global Warming Report Warns: Seas Will Rise." *St. Petersburg Times*, October 24, 2001, 3-B.

Poggioli, Sylvia. "Venice Struggling with Increased Flooding." National Public Radio, *Morning Edition*, November 29, 2002. (Lexis).

Pohl, Otto. "New Jellyfish Problem Means Jellyfish Are Not the Only Problem." *New York Times*, May 21, 2002, F-3.

Quadfasel, Detlef. "Oceanography: The Atlantic Heat Conveyor Slows." *Nature* 438 (December 1, 2005): 565–566.

Radford, Tim. "As the World Gets Hotter, Will Britain Get Colder? Plunging Temperatures Feared after Scientists Find Gulf Stream Changes." *London Guardian*, June 21, 2001, 3.

———. "Coral Reefs Face Total Destruction within 50 Years." *London Guardian*, September 6, 2001, 9.

———. "Ten Key Coral Reefs Shelter Much of Sea Life: American Association Scientists Identify Vulnerable Marine 'Hot Spots' with the Richest Biodiversity on Earth." *London Guardian*, February 15, 2002, 12.

———. "2020: The Drowned World." *London Guardian*, September 11, 2004, 10.

Radowitz, John von. "Global Warming 'Smoking Gun' Found in the Oceans." Press Associated Ltd., February 18, 2005. (Lexis).

Rahmstorf, Stefan. "Thermohaline Circulation: The Current Climate." *Nature* 421 (February 13, 2003): 699.

"Report Says Oceans Hit by Carbon Dioxide Use." *Boston Globe* in *Omaha World-Herald*, July 17, 2004, 5-A.

Reuters in *South China Morning Post* [No Title], March 15, 2001, 11. (Lexis).

Revkin, Andrew C. "Two New Studies Tie Rise in Ocean Heat to Greenhouse Gases." *New York Times*, April 13, 2001, A-15.

Richardson, Franci. "Sharks Take the Bait: Experts: Sightings in Maine an 'Unusual Circumstance.'" *Boston Herald*, August 11, 2002, 3.

Rickaby, R.E.M., and P. Halloran. "Cool El Niño during the Warmth of the Pliocene?" *Science* 307 (March 25, 2005): 1948–1952.

"Rising Seas Threaten Bay Marshes." Environment News Service, April 11, 2002. http://ens-news.com/ens/apr2002/2002L-04-11-09.html.

"Rising Tide: Who Needs Essex Anyway." *London Guardian*, June 12, 2003, 4.

Roberts, Callum M., Colin J. McClean, John E. N. Veron, Julie P. Hawkins, Gerald R. Allen, Don E. McAllister, et al. "Marine Biodiversity Hotspots and Conservation Priorities for Tropical Reefs." *Science* 295 (February 15, 2002): 1280–1284.

Roberts, Greg. "Great Barrier Grief as Warm-Water Bleaching Lingers." *Sydney Morning Herald*, January 20, 2003, 4.

Rowan, Rob. "Thermal Adaptation in Reef Coral Symbionts." *Nature* 430 (August 12, 2004): 742.

Rubin, Daniel. "Venice Sinks as Adriatic Rises." Knight-Ridder News Service, July 1, 2003. (Lexis).

Sadler, Richard, and Geoffrey Lean. "North Sea Faces Collapse of Its Ecosystem." *London Independent*, October 19, 2003, 12.

Savill, Richard. "Tropical Fish Hooked on Channel Holidays." *London Daily Telegraph*, August 12, 2004, 7.

Schiermeier, Quirin. "Researchers Seek to Turn the Tide on Problem of Acid Seas." *Nature* 430 (August 19, 2004): 820.

Schleifstein, Mark. "The Gulf [of Mexico] Will Rise, Report Predicts: Maybe 44 Inches, Scientists Say." *New Orleans Times-Picayune*, October 24, 2001, 1.

Schmittner, Andreas. "Decline of the Marine Ecosystem Caused by a Reduction in the Atlantic Overturning Circulation." *Nature* 434 (March 31, 2005): 628–633.

Seager, R., D. S. Battisti, J. Yin, N. Gordon, N. Naik, A. C. Clement, et al. "Is the Gulf Stream Responsible for Europe's Mild Winters?" *Quarterly Journal of the Royal Meteorological Society* 128 (2002): 2563–2586.

Serreze, M. C., J. E. Walsh, F. C. Chapin, T. Osterkamp, M. Dyurgerov, V. Romanovsky, et al. "Observational Evidence of Recent Change in the Northern High-Latitude Environment." *Climatic Change* 46 (2000): 159–207.

"70 Cities in Indonesia Will Be Inundated." Antara, the Indonesian National News Agency, September 25, 2002. (Lexis).

"Shanghai Mulls Building Dam to Ward Off Rising Sea Levels." Agence France Presse, February 9, 2004. (Lexis).

"Sharks in Alaskan Waters Could Herald Global Warming." Environment News Service, February 19, 2002. http://ens-news.com/ens/feb2002/2002L-02-19-09.html.

Sheppard, Charles R. C. "Predicted Recurrences of Mass Coral Mortality in the Indian Ocean." *Nature* 425 (September 18, 2003): 294–297.

Siggins, Lorna. "Warm-Water Anchovies Landed by Trawlers in Donegal Bay." *Irish Times*, December 12, 2001, 1.

"Slowing Ocean Currents Could Freeze Europe." Environment News Service, February 21, 2002. http://ens-news.com/ens/feb2002/2002L-02-21-09.html.

Smith, Craig S. "One Hundred and Fifty Nations Start Groundwork for Global Warming Policies." *New York Times*, January 18, 2001, 7.

Smith, Graeme. "Fishermen Fear the Worst over Cod: Parallels with Canada Collapse When 30,000 Jobs Were Lost." *Glasgow Herald*, October 23, 2002, 7.

Spalding, Mark. "Coral Grief: Rising Temperatures, Pollution, Tourism and Fishing Have All Helped to Kill Vast Stretches of Reef in the Indian Ocean. Yet, with Simple Management, Says Mark Spalding, the Marine Life Can Recover." *London Guardian*, September 12, 2001, 8.

———. *World Atlas of Coral Reefs*. Berkeley, University of California Press, 2001.

Speth, James Gustave. *Red Sky at Morning: America and the Crisis of the Global Environment*. New Haven, CT: Yale University Press, 2004.

Stiffler, Linda, and Robert McClure. "Effects Could Be Profound." *Seattle Post-Intelligencer*, November 13, 2003, A-8.

Stocker, Thomas F. "Global Change: South Dials North." *Nature* 424 (July 31, 2003): 496–499.

———, Reto Knutti, and Gian-Kasper Plattner. "The Future of the Thermo-haline Circulation—A Perspective." In *The Oceans and Rapid Climate Change: Past, Present, and Future*, ed. Dan Seidov, Bernd J. Haupt, and Mark Maslin, 277–293. Washington, D.C.: American Geophysical Union, 2001.

Stoddard, Ed. "Global Warming Threatens 'Living Fossil' Fish: Coelacanths Have Existed 400 Million Years." Reuters in *Ottawa Citizen*, July 14, 2001, B-4.

"Study Reports Large-Scale Salinity Changes in Oceans: Saltier Tropical Oceans, Fresher Ocean Waters Near Poles Are Further Signs of Global Warming's Impacts on Planet." AScribe Newswire, December 17, 2003. (Lexis).

"Study Reveals Increased River Discharge to Arctic Ocean: Finding Could Mean Big Changes to Global Climate." AScribe Newswire, December 12, 2002. (Lexis).

Taalas, P., J. Damski, E. Kyro, M. Ginzburg, and G. Talamoni. "The Effect of Stratospheric Ozone Variations on UV Radiation and on Tropospheric Ozone at High Latitudes." *Journal of Geophysical Research* 102:D1 (1997): 1533–1543.

———, E. Kyrö, K. Jokela, T. Koskela, J. Damski, M. Rummukainen, et al. "Stratospheric Ozone Depletion and Its Impact on UV Radiation and on Human Health." *Geophysica* 32 (1996): 127–165.

Thompson David W. J., and J. M. Wallace. "The Arctic Oscillation Signature in the Wintertime Geopotential Height and Temperature Fields." *Geophysical Research Letters* 25 (1998): 1297–1300.

Thorpe, R. B., J. M. Gregory, T. C. Johns, R. A. Wood, and J. F. B. Mitchell. "Mechanisms Determining the Atlantic Thermohaline Circulation Response to Greenhouse Gas Forcing in a Non-Flux-Adjusted Coupled Climate Model." *Journal of Climate* 14 (July 15, 2001): 3102–3116.

Toner, Mike. "Microscopic Ocean Life in Global Decline: Temperature Shifts a Cause or an Effect?" *Atlanta Journal and Constitution*, August 9, 2002, 3-A.

———. "Oceans' Acidity Worries Experts: Report: Carbon Dioxide on Rise, Marine Life at Risk." *Atlanta Journal and Constitution*, September 25, 2003. (Lexis).

Trenberth, Kevin E., and Timothy J. Hoar. "The 1990–1995 El Niño-Southern Oscillation Event: Longest on Record." *Geophysical Research Letters* 23 (1) (January 1, 1996): 57–60.

Unwin, Brian. "Tropical Birds and Exotic Sea Creatures Warm to Britain's Welcoming Waters." *London Independent*, August 20, 2001, 7.

Urban, Frank E., Julia E. Cole, and Jonathan T. Overpeck. "Influence on Mean Climate Variability from a 155-Year Tropical Pacific Coral Record." *Nature* 407 (October 26, 2000): 989–993.

Verburg, Piet, Robert E. Hecky, and Hedy Kling. "Ecological Consequences of a Century of Warming in Lake Tanganyika." *Science* 301 (July 25, 2003): 505–507.

Verschuren, Dirk. "Global Change: The Heat on Lake Tanganyika." *Nature* 424 (August 14, 2003): 731–732.

Walther, Gian-Reto, Eric Post, Peter Convey, Annette Menzel, Camille Parmesan, Trevor J. C. Beebee, et al. "Ecological Responses to Recent Climate Change." *Nature* 416 (March 28, 2002): 389–395.

Wara, Michael W., Ana Christina Ravelo, and Margaret L. Delaney. "Permanent El Niño-Like Conditions during the Pliocene Warm Period." *Science* 309 (July 29, 2005): 758–761.

"Warming Could Submerge Three of India's Largest Cities: Scientist." Agence France Presse, December 6, 2003. (Lexis).

"Warming Doom for Great Barrier Reef." Australian Associated Press in *Hobart (Australia) Mercury*, February 16, 2002. (Lexis).

"Warming Streams Could Wipe Out Salmon, Trout." Environment News Service, May 22, 2002. http://ens-news.com/ens/may2002/2002L-05-22-06.html.

Watson, Jeremy. "Plan to Hold Back Tides of Venice Runs into Flood of Opposition from Greens." *Scotland on Sunday*, December 30, 2001, 18.

Weart, Spencer R. *The Discovery of Global Warming*. Cambridge, MA: Harvard University Press, 2003.

Webster, P. J., G. J. Holland, J. A. Curry, and H.-R. Chang. "Changes in Tropical Cyclone Number, Duration, and Intensity in a Warming Environment." *Science* 309 (September 16, 2005): 1844–1846.

Weller, G. "Regional Impacts of Climate Change in the Arctic and Antarctic." *Annals of Glaciology* 27 (1998): 543–552.

Whoriskey, Peter, and Joby Warrick. "Report Revises Katrina's Force: Hurricane Center Downgrades Storm to Category 3 Strength." *Washington Post*, December 22, 2005, A-3. www.washingtonpost.com/wp-dyn/content/article/2005/12/21/AR2005122101960.html.

Williams, Brian. "Reef Down to Half Its Former Self." *Queensland (Australia) Courier Mail*, June 24, 2004, 11.

Wilson, Paul A., and Richard D. Norris. "Warm Tropical Ocean Surface and Global Anoxia during the Mid-Cretaceous Period." *Nature* 412 (July 26, 2001): 425–429.

Woodcock, John. "Coral Reefs at Risk from Man-Made Cocktail of Poisons." *The Scotsman*, September 11, 2001, 4.

Zachos, James C., Ursula Röhl, Stephen A. Schellenberg, Appy Sluijs, David A. Hodell, Daniel C. Kelly, et al. "Rapid Acidification of the Ocean during the Paleocene-Eocene Thermal Maximum." *Science* 308 (June 10, 2005): 1611–1615.

Zhang, Keqi, Bruce C. Douglas, and Stephen P. Leatherman. "Global Warming and Coastal Erosion." *Climatic Change* 64 (1/2) (May 2004): 41–58.

DATE			